iOS 项目开发教程

郭朗　卓国锋 / 主编

孟瑞　刘盼盼 / 编著

清华大学出版社

北京

内 容 简 介

本书基于 Apple 最新发布的 iOS 7 编写。书中循序渐进地介绍了 iOS 项目开发的一般步骤和基础知识，主要内容包括设计及美化用户界面，用 Interface Builder 构建视图，切换和弹出框，用导航控制器导航页面，用表视图结构化数据，读写和显示数据以及地图和定位功能等。

本书通过简洁的语言和详细的步骤，帮助读者迅速掌握开发 iOS 应用程序所需的基本知识，适合有一定编程经验的读者阅读。本书可作为高等学校教材，也可供从事 iOS 开发的人员参考。

图书在版编目(CIP)数据

iOS 项目开发教程/郭朗，卓国锋主编. —北京：清华大学出版社，2018（2022.3重印）
ISBN 978-7-302-50901-1

Ⅰ. ①i… Ⅱ. ①郭… ②卓… Ⅲ. ①移动终端－应用程序－程序设计－教材 Ⅳ. ①TN929.53

中国版本图书馆 CIP 数据核字(2018)第 185376 号

责任编辑：焦 虹 李 晔
封面设计：常雪影
责任校对：白 蕾
责任印制：曹婉颖

出版发行：清华大学出版社
网 址：http://www.tup.com.cn，http://www.wqbook.com
地 址：北京清华大学学研大厦 A 座 **邮 编**：100084
社 总 机：010-83470000 **邮 购**：010-62786544
投稿与读者服务：010-62776969，c-service@tup.tsinghua.edu.cn
质量反馈：010-62772015，zhiliang@tup.tsinghua.edu.cn
课件下载：http://www.tup.com.cn，010-62795954
印 装 者：天津鑫丰华印务有限公司
经 销：全国新华书店
开 本：185mm×260mm **印 张**：15.75 **字 数**：366 千字
版 次：2018 年 9 月第 1 版 **印 次**：2022 年 3 月第 2 次印刷
定 价：48.00 元

产品编号：062178-02

前言 Foreword

iOS 平台改变了公众对移动设备的看法，iPhone 凭借着功能齐备的应用程序、界面架构以及其他平台无法媲美的触控，提供了方便的手机功能，证明了小屏幕也能成为高效的工作区。

在 Apple 手机的开发人员看来，用户体验至关重要。他们设计了 iOS，其外观和行为不再像笨拙的桌面应用程序的移植版本，用户能够使用手指(而不是光笔或键盘)控制手机，从而让应用程序使用起来更加自然而有趣。

通过 AppStore，Apple 公司向开发人员提供了一种最佳的数字发布系统。开发人员可以将应用程序提交到 AppStore，且只需要支付少量的年度开发人员会费。人们开发了针对各种领域的游戏和应用程序，其范围涵盖了从学前教育到退休生活的所有阶段。由于 iPhone、iPod Touch 和 iPad 的用户群非常庞大，因此不管什么内容都能找到合适的用户。

Apple 公司每年都会发布新的 iOS 设备，其速度更快、分辨率更高、功能更强大。每次硬件更新都带来了新的开发机会，提供了将艺术融合到软件中的新途径。

1. 谁能成为 iOS 开发人员

只要有学习兴趣，有时间探索和使用 Apple 开发工具，并拥有一台运行 Lion 的 Inter Macintosh 计算机，便可开始 iOS 开发了。虽然不可能在一夜之间就开发出 iOS 应用程序，但只要多加练习，完全可以在几天内编写出第一款应用程序。在 Apple 开发工具上花费的时间越多，创建出激动人心的应用程序的可能性就越大。

2. 本书适合的读者

本书是为熟悉 iOS 开发语言 Objective-C 的读者编写的，读者不需要具有 Cocoa 和 Apple 开发的经验；当然，读者如果有一定的开发经验，将会更容易掌握这些工具和技术。尽管如此，本书对读者还是有一定要求的，就是读者必须愿意花费时间进行学习与练习。如果读者只是阅读每章的内容，而不完成其中的项目，就会错过一些重要概念。另外，读者还需认真阅读 Apple 开发文档，并研究本书介绍的主题，这样才能掌握全部内容。

编著者

目录 Contents

第1章 chapter 1

iOS应用开发概述

本章首先介绍 iOS 应用开发历史与现状，然后对 iOS 框架进行整体介绍，最后介绍苹果开发工具包 iOS SDK 及开发环境的搭建。

1.1 iOS应用开发的历史与现状

苹果 iOS 是由苹果公司开发的手持设备操作系统。苹果公司最早于 2007 年 1 月 9 日的 Macworld 大会上公布这个了系统，最初是设计给 iPhone 使用的，后来陆续套用到 iPod touch、iPad 以及 AppleTV 等苹果产品上。iOS 与苹果的 Mac OS X 操作系统一样，也是以 Darwin 为基础的，因此同样属于类 UNIX 的商业操作系统。

1. 历史

2007 年 10 月 17 日，苹果公司发布了第一个本地化 iPhone 应用程序开发包(SDK)。2008 年 3 月 6 日，苹果公司发布了第一个测试版开发包，并且将 iPhone runs OSX 改名为 iPhoneOS。2010 年 2 月 27 日，苹果公司发布了 iPad，iPad 同样搭载了 iPhoneOS。同年，苹果公司重新设计了 iPhoneOS 的系统结构和自带程序。2010 年 6 月，苹果公司将 iPhoneOS 改名为 iOS，同时还获得了思科 iOS 的名称授权。2012 年 6 月，苹果公司在 WWDC 2012 上宣布了 iOS6，提供了超过 200 项新功能。2013 年 6 月 10 日，苹果公司在 WWDC 2013 上发布了 iOS7，几乎重绘了所有的系统 App，去掉了所有的仿实物化，整体设计风格转为扁平化设计。2014 年 6 月 3 日，苹果公司在 WWDC 2014 上发布了 iOS8，并提供了开发者预览版更新。

2. 现状

iPhone 在全球创造的庞大应用市场，使应用开发公司开始争抢 iOS 软件开发人才。另外，由于 iOS 系统开发技术位于全球手机系统的前端，其他系统平台应用开发公司和系统研发公司同时也在高薪招聘。72%的招聘公司正在招聘 iOS 平台开发人才，其中 38%的招聘公司表示，iOS 平台开发经验要比任何其他平台开发经验更受招聘公司的青睐。由于国内 iOS 软件开发起步相对较晚，人才培养机制更是远远跟不上市场的发展速度，有限的 iOS 开发人才成了国内企业必争的资源，甚至有的企业不得不考虑通过收购来填补人才空缺。一名 iOS 开发新手要比普通软件开发新手高出约 20%～30%的薪资，符合条件或有项目经验的开发工程师更是有价无市。

1.2 iOS 应用的基本架构

iOS 的基本架构分为四个层次：核心操作系统层(Core OS layer)、核心服务层(Core Services layer)、媒体层(Media layer)和可触摸层(Cocoa Touch layer)，如图 1.1 所示。

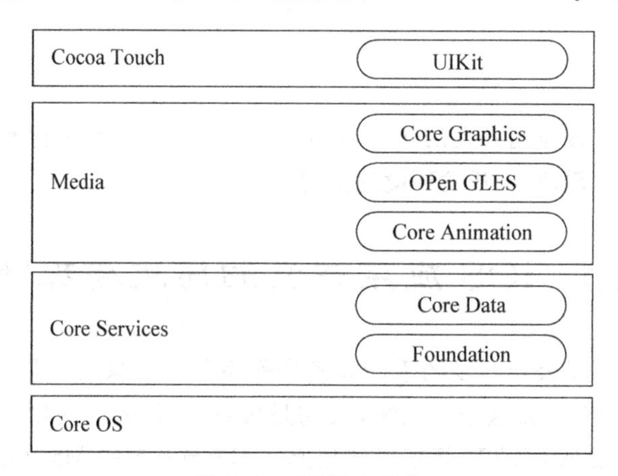

图 1.1 iOS 基本架构

1.2.1 可触摸层

可触摸层包含创建 iOS 应用程序所需的关键框架。上至实现应用程序可视界面，下至与高级系统服务交互，都需要该层技术提供底层基础。在开发应用程序的时候，尽可能不要使用更底层的框架，而是使用该层的框架。该框架中比较重要的框架如下。

1. Address Book UI 框架

Address Book UI 框架(AddressBookUI. framework)是一套 Objective-C 的编程接口，能够显示创建或者编辑联系人的标准系统界面。该框架简化了应用程序显示联系人信息的工作，另外它也可以确保应用程序使用的界面和其他应用程序相同，进而保证跨平台的一致性。

2. Game Kit 框架

iOS 3.0 引入了 Game Kit 框架(GameKit. framework)。该框架支持点对点连接及游戏内语音功能，可以通过该框架为应用程序增加点对点网络功能。此框架通过一组建构于 Bonjour 之上的简单而强大的类提供网络功能，这些类将许多网络细节抽象出来，从而让没有网络编程经验的开发者可以更加容易地将网络功能整合到应用程序中。

3. Map Kit 框架

iOS 3.0 导入了 Map Kit 框架(MapKit.framework),该框架提供一个可被嵌入到应用程序的地图界面,该界面包含一个可以滚动的地图视图。开发者可以在视图中添加定制信息,并可将其嵌入到应用程序视图,通过编程的方式设置地图的各种属性(包括当前地图显示的区域以及用户的方位),也可以使用定制标注或标准标注(例如使用测针标记)突出显示地图中的某些区域或额外的信息。

4. UIKit 框架

UIKit 框架(UIKit.framework)的 Objective-C 编程接口为实现 iOS 应用程序的图形及事件驱动提供关键基础。

1.2.2 媒体层

媒体层包含图形技术、音频技术和视频技术,这些技术相互结合就可为移动设备带来最好的多媒体体验。更重要的是,它们让创建外观音效俱佳的应用程序变得更加容易。开发者可以使用 iOS 的高级框架更快速地创建高级的图形和动画,也可以通过底层框架访问必要的工具,从而以特定方式完成某种任务。该框架中比较重要的框架如下。

1. AV Foundation 框架

iOS 2.2 引入了 AV Foundation 框架(AVFoundation.framework),该框架包含的 Objective-C 类可用于播放音频内容。通过使用该框架,开发者可以播放声音文件或播放内存中的音频数据,也可以同时播放多个声音,并对各个声音的播放特效进行控制。

2. Core Audio 框架

Core Audio 框架提供 C 语言接口,可用于操作立体声音频。通过 iOS 系统 Core Audio 框架,开发者可以在应用程序中生成、录制、混合或播放音频,也可通过该框架访问设备的震动功能(对于支持震动功能的设备)。

3. Core Graphics 框架

Core Graphics 框架(CoreGraphics.framework)包含 Quartz 2D 绘图 API 接口。Quartz 是 Mac OS X 系统使用的向量绘图引擎,它支持基于路径绘图,抗锯齿渲染,渐变,图片,颜色,坐标空间转换,PDF 文件的创建、显示和解析。虽然 API 基于 C 语言,但是它使用基于对象的抽象以表示基本绘图对象,这样可以让开发者更方便地保存并复用图像内容。

4. Core Text 框架

iOS 3.2 引入了 Core Text 框架(CoreText.framework),该框架包含一组简单高效

的C接口,可用于对文本进行布局以及对字体进行处理。Core Text框架提供一个完整的文本布局引擎,开发者可以通过它管理文本在屏幕上的摆放。所管理的文本也可以使用不同的字体和渲染属性。

1.2.3 核心服务层

核心服务层为所有的应用程序提供基础系统服务。可能应用程序并不直接使用这些服务,但它们是系统许多部分赖以建构的基础。该框架中比较重要的框架如下。

1. Address Book 框架

Address Book框架(AddressBook. framework)支持编程访问存储于用户设备中的联系人信息。如果应用程序要使用联系人信息,则可通过该框架访问并修改用户联系人数据库的记录。例如,通过使用该框架,聊天程序可以获取一个联系人列表,利用此列表初始化聊天会话,并在联系人视图显示列表中的联系人。

2. Core Data 框架

iOS 3.0引入Core Data框架(CoreData. framework)。Core Data框架是一种管理"模型-视图-控制器"应用程序数据模型的技术,适用于数据模型已经高度结构化的应用程序。利用此框架,开发者再也不需要通过编程定义数据结构,而是通过Xcode提供的图形工具构造一份代表数据模型的图表。在程序运行的时候,Core Data框架就会创建并管理数据模型的实例,同时还可对外提供数据模型访问接口。

3. Core Foundation 框架

Core Foundation框架(CoreFoundation. framework)是一组C语言接口,为iOS应用程序提供基本数据管理和服务功能。

4. Core Location 框架

Core Location框架(CoreLocation. framework)可用于定位某个设备的当前经纬度。它可以利用设备具备的硬件,通过附近的GPS、蜂窝基站或者WiFi信号等信息计算用户方位。Maps应用程序就是利用此功能在地图上显示用户当前位置。开发者可将此技术结合到应用程序,以此向用户提供方位信息。例如,应用程序可根据用户当前位置搜索附近饭店、商店或其他设施。

5. Core Media 框架

iOS 4.0引入了Core Media框架(CoreMedia. framework)。此框架提供AV Foundation框架使用的底层媒体类型。只有少数需要对音频或视频创建及展示进行精确控制的应用才会涉及该框架,其他大部分应用程序都用不上它。

1.2.4 核心操作系统层

核心操作系统层的底层功能是很多其他技术的构建基础。通常情况下，这些功能不会直接应用于应用程序，而是应用于其他框架，但是在直接处理安全事务或与某个外设通信的时候，必须应用到该层的框架。该框架中比较重要的框架如下。

1. Accelerate 框架

iOS 4.0 引入了 Accelerate 框架(Accelerate.framework)。该框架的接口可用于执行数学以及 DSP 运算。和开发者个人编写的库相比，该框架的优点在于它根据现存的各种 iOS 设备的硬件配置进行过优化，因此开发者只需一次编码就可确保它在所有设备都能高效运行。

2. External Accessory 框架

iOS 3.0 引入了 External Accessory 框架(ExternalAccessory.framework)，通过它来支持 iOS 设备与绑定光盘通信。光盘可以通过一个 30 针的基座接口和设备相连，也可通过蓝牙连接。通过 External Accessory 框架，开发者可以获得每个外设的信息并初始化一个通信会话。通信会话初始化完成之后，可以使用设备支持的命令直接对其进行操作。

3. Security 框架

iOS 系统不但提供内建的安全功能，还提供 Security 框架(Security.framework)用于保证应用程序所管理的数据的安全。该框架提供的接口可用于管理证书、公钥、私钥以及信任策略。它支持生成加密的安全伪随机数。同时，它也支持对证书和 Keychain 密钥进行保存，是用户敏感数据的安全仓库。

CommonCrypto 接口还支持对称加密、HMAC 以及 Digests。实际上，Digests 的功能和 OpenSSL 库常用的功能兼容，只是 iOS 无法使用 OpenSSL 库。

4. System 框架

系统层包括内核环境、驱动及操作系统底层 UNIX 接口。内核以 Mach 为基础，它负责操作系统的各个方面，包括管理系统的虚拟内存、线程、文件系统、网络以及进程间通信。这一层包含的驱动是系统硬件和系统框架的接口。出于安全方面的考虑，内核和驱动只允许少数系统框架和应用程序访问。

1.3 iOS 应用开发的特点

iOS 平台上的程序拥有一些共性，这些特性会影响自身的使用体验。与这些特性相适应的程序会更加成功，与设备一起为用户提供更好的使用体验。

1. 最先响应屏幕

iOS对屏幕反应的优先级是最高的，它的响应顺序依次为Touch→Media→Service→Core架构。换句话说，用户触摸屏幕之后，系统会最优先处理屏幕显示，然后才是媒体(Media)、服务(Service)以及Core架构。因此iOS应用很少出现卡顿或者延迟现象。

2. 基于GPU加速

iOS系统对图形的各种特效处理基本上都是基于GPU硬件进行加速的，可以不用完全借助CPU或者程序本身，而是通过GPU进行渲染以达到更流畅的操控表现，使得iOS应用在操控过程中有着非常好的流畅性。

3. 机制效率高

iOS采用Objective-C作为开发语言，编译器为LLVM。LLVM编译后的代码又被苹果专为iOS架构进行优化，运行过程中不需要虚拟机插手，执行效率非常高，并且系统不需要占用大量内存来换取执行速度，从而使得程序占用系统资源非常少，显著提高了程序运行的流畅度。

4. 硬件利用率高

iOS平台非常封闭，所有的App运行对象都比较单一。由于每个应用程序都运行在iPhone、iPad等iOS产品中，所以有着很高的硬件利用效率。iOS应用开发也因为iOS平台软硬件垂直整合而受益，使得开发者很容易针对iOS设备进行适配，从而极大地降低了开发成本和周期。

1.4 iOS开发工具简介

打开Mac上的App Store，搜索并下载Xcode。

下载前需注册iOS开发者账号，登录后即可下载，大小约为3.5GB，包括Xcode、Interface Builder和模拟器等工具，如图1.2所示。

1. Xcode

要开发iOS应用，需在Mac OS X上运行Xcode开发工具。Xcode是Apple的开发工具套件，支持项目管理、编辑代码、构建可执行程序、代码级调试、代码的版本管理、性能调优等。这个套件的核心是Xcode应用本身，它提供了基本的源代码开发环境，如图1.3所示。

2. Interface Builder

利用Interface Builder，可以拖曳需要的组件在程序窗口上进行装配。组件中包含标

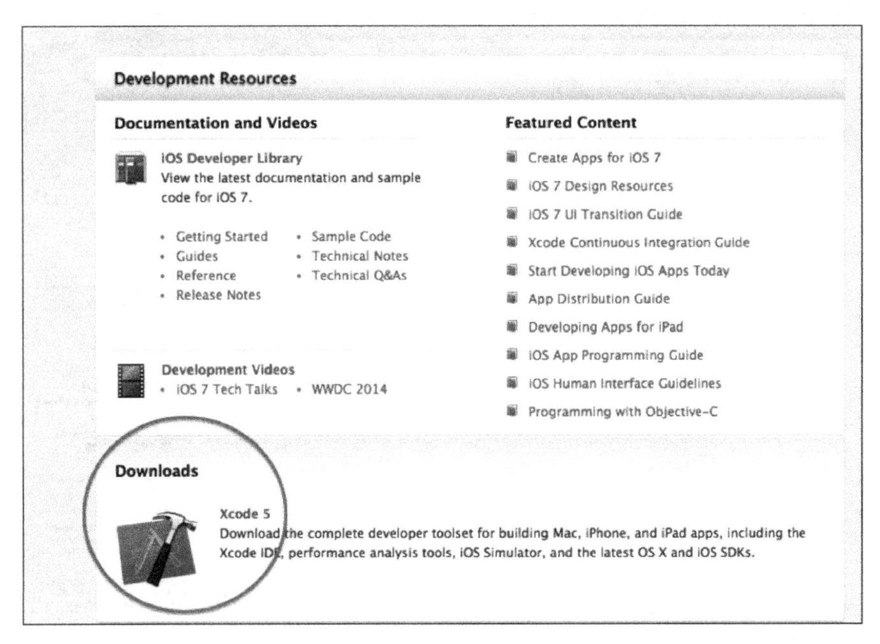

图 1.2　下载开发工具

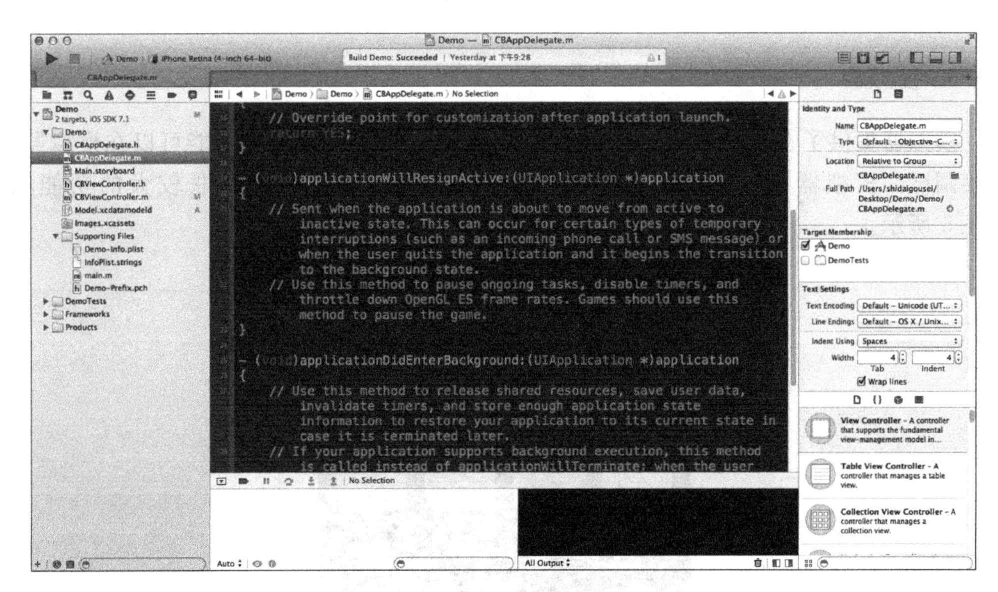

图 1.3　Xcode 界面

准的系统控件,如开关(switches)、文本框和按钮,还有定制的视图来表示程序提供的视图。在窗口表面上放置组件之后,拖曳它们可以确定位置。使用观察器(inspector)设置其属性,建立这些对象和代码之间的联系。最后将内容保存在一个 xib 文件中,这是一个自定义的资源文件格式,如图 1.4 所示。

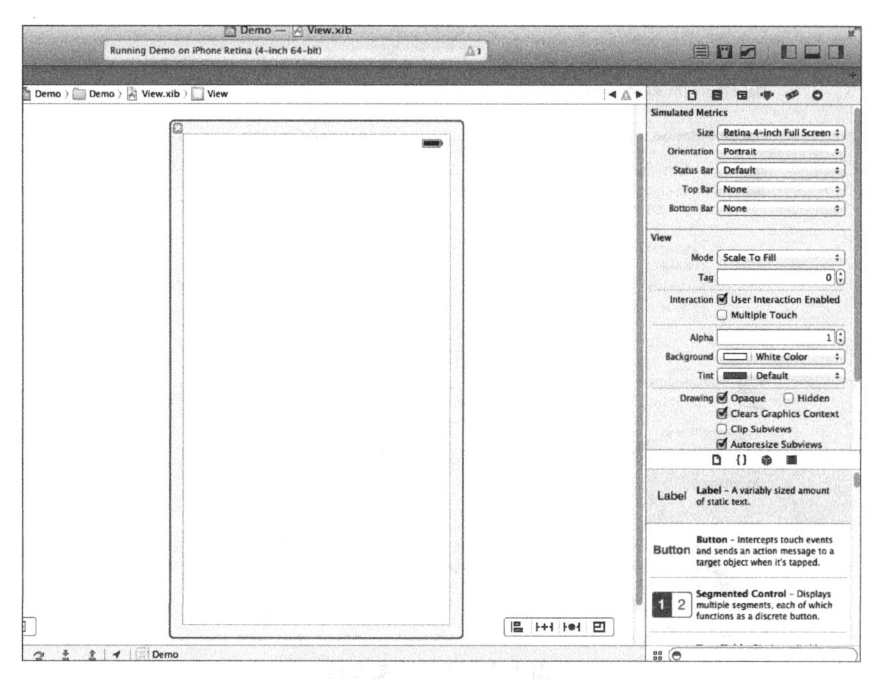

图 1.4 Interface Builder 界面

3. 模拟器

模拟器提供了在苹果电脑上开发 iOS 产品时的虚拟设备，部分功能可以在模拟器上直接调试。模拟器不支持 GPS 定位、摄像头、指南针等与硬件关联较大的功能，如图 1.5 所示。

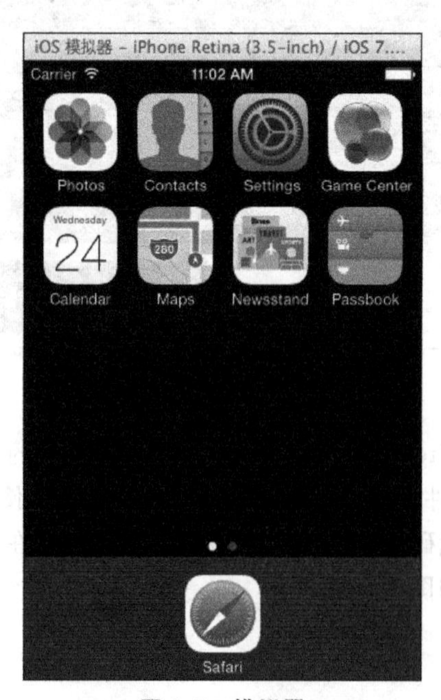

图 1.5 模拟器

1.5 创建并运行第一个iOS应用

本节将演示如何用Xcode创建一个新项目。

(1) 双击Xcode图标,启动Xcode,如图1.6所示。

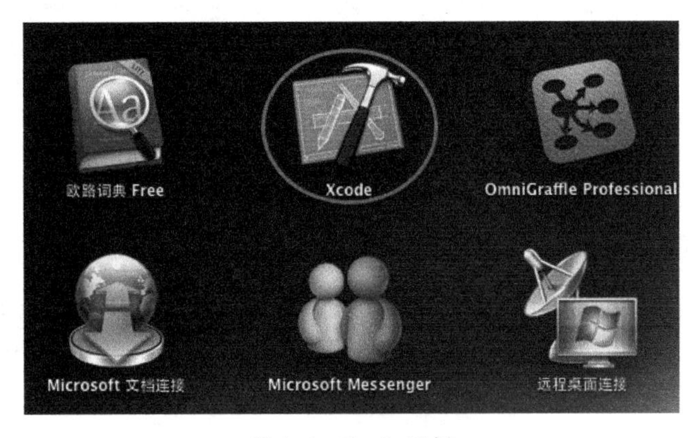

图1.6 Xcode图标

(2) 单击Create a new Xcode project图标,创建一个新工程项目,如图1.7所示。

图1.7 创建新工程项目

(3) 选择iOS项目模板中的Single View Application图标,单击Next按钮,如图1.8所示。

(4) 依次输入项目名称、机构名、公司标识、类名前缀后,单击Next按钮,如图1.9所示。

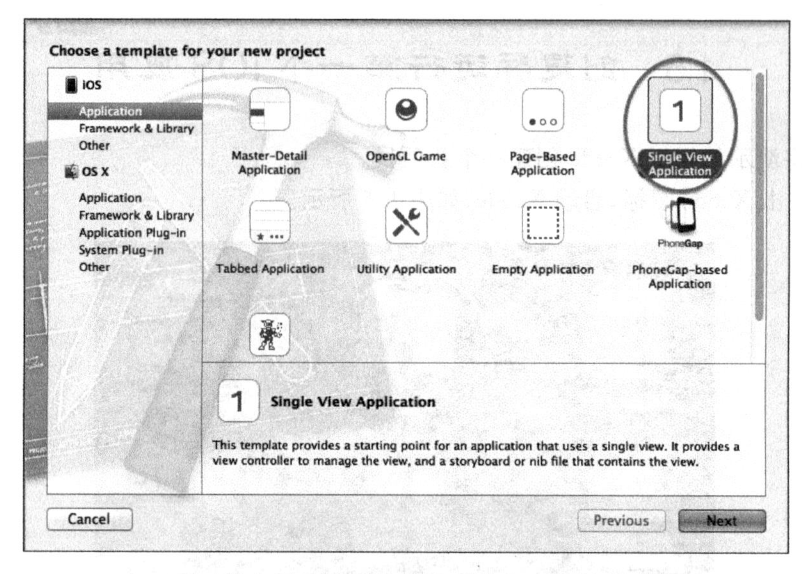

图 1.8　选择 Single View Application

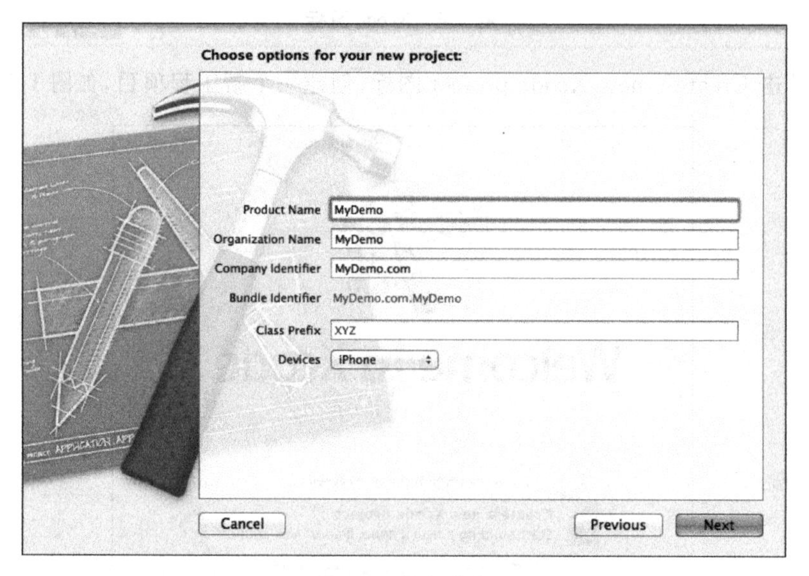

图 1.9　输入内容

(5) 选择项目保存的目录，单击 Create，如图 1.10 所示。

(6) 单击图 1.10 的界面左上角的▶按钮，运行程序。Xcode 会自动启动 iPhone 模拟器，如图 1.11 所示。

这样，第一个 iOS 应用就运行成功了。

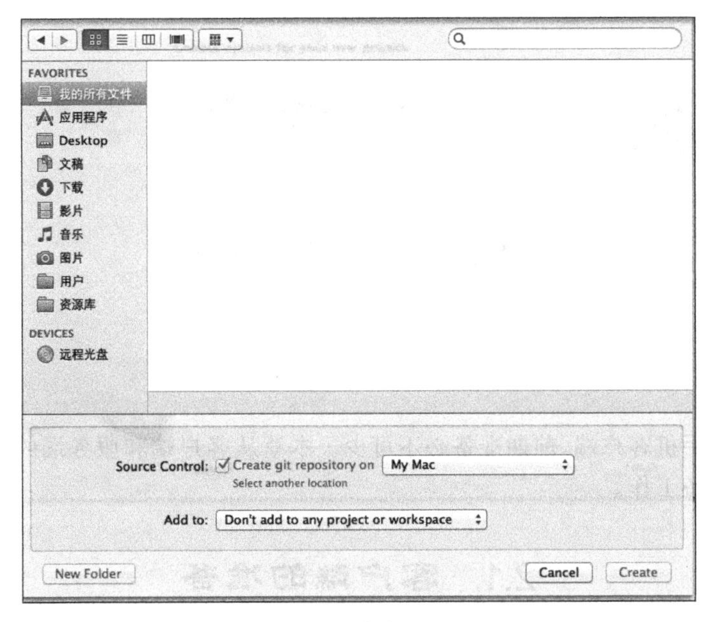

图 1.10 单击 Create

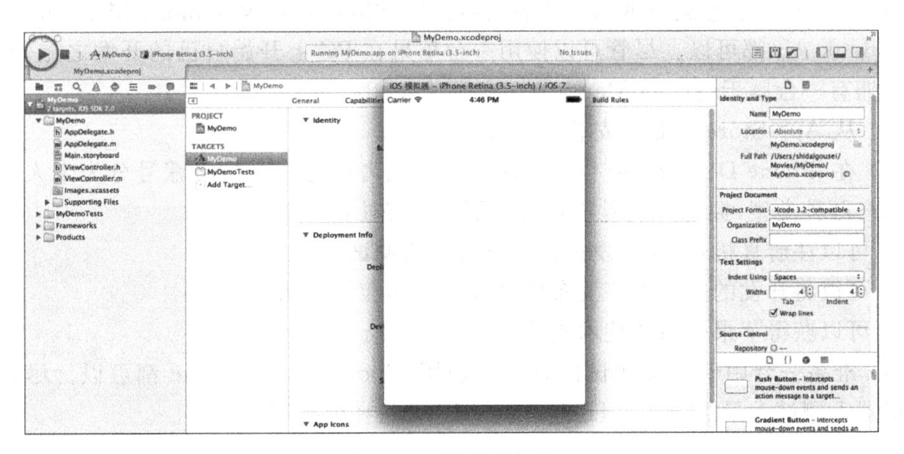

图 1.11 编译并运行

1.6 基础知识与技能回顾

本章简要介绍了 iOS 系统的发展历程和 iOS 平台的架构特性，详细列出了 Mac 系统下 iOS 开发平台环境的搭建步骤，为后续 iOS 项目开发的学习打下了基础。

练　　习

项目

功能描述：创建并运行第一个 iOS 应用。

第2章 chapter 2

为开发做好准备

开发一款手机客户端，前期准备必不可少。本章从客户端和服务端的角度分别介绍前期开发的准备工作。

2.1 客户端的准备

(1) 准备一台能够运行 Mac OS 的机器。MacBook Pro、Mac mini、iMac、Mac Pro 甚至 MacBook Air 都可以。尽管可以使用一台苹果在 PC 上开始自己的开发之旅，不过还是建议拥有一台属于自己的 Mac。

(2) 从 App Store 下载并安装 Xcode。

(3) 在 iPhone Dev Center 付费注册一个开发者账号。开发者账号分为个人用户和企业用户两种类型，拥有开发者账号有以下三点好处：

- 可以注册真机，并在开发过程中进行真机调试。
- 可以将开发的应用程序发布到 App Store(仅限个人用户)。
- 可以获得苹果公司的技术支持。

(4) 准备一台用于开发测试的真机。使用 iPhone、iTouch、iPad 都可以。尽管能够在模拟器上测试大量的 API，但还是有些 API 无法在模拟器上工作。

2.2 服务端的准备

本项目服务端采用 Tomcat＋MySQL 方式。由于服务端不是本书重点，因此简要介绍。按照下面的步骤配置安装，完成本地服务端的搭建。

2.2.1 Web Services 的搭建

什么是 Tomcat? Tomcat 是由 Apache 软件基金会下属的 Jakarta 项目开发的一个 Servlet 容器，按照 Sun Microsystems 提供的技术规范，实现了对 Servlet 和 JavaServer Page(JSP)的支持，并提供了作为 Web 服务端的一些特有功能，如 Tomcat 管理和控制平台、安全域管理和 Tomcat 阀等。由于 Tomcat 本身内含了一个 HTTP 服务端，因此它也可以被视作一个单独的 Web 服务端。

1. 下载安装 Tomcat

(1) 访问 http://tomcat.apache.org/download-70.cgi，选择具体文件进行下载，如图 2.1 所示。

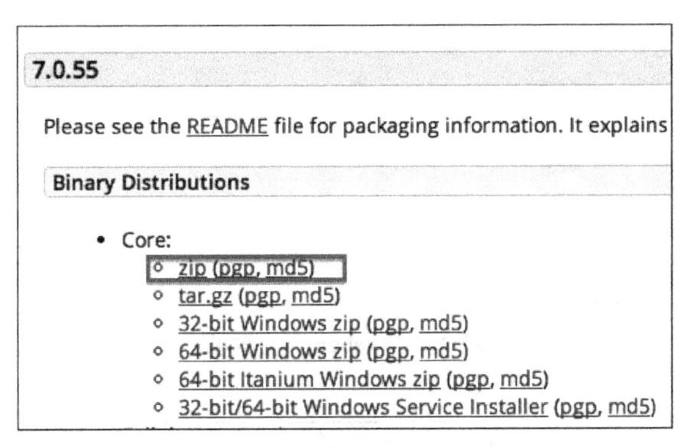

图 2.1 下载 Tomcat

(2) 解压下载文件到目录/Library 中，并把文件夹名改为 Tomcat。

(3) 打开终端，输入命令：

```
sudo chmod 755 /Library/Tomcat/bin/* .sh
```

按回车键之后会提示输入密码，请输入管理员密码，之后输入下面的命令并回车。

```
sudo sh /Library/Tomcat/bin/startup.sh
```

(4) 打开浏览器，输入 http://localhost:8080/，回车之后如果看到 Apache Tomcat，则表示已经成功运行 Tomcat，如图 2.2 所示。在终端输入命令 sudo sh /Library/Tomcat/bin/shutdown.sh 并回车，可以关闭 Tomcat。

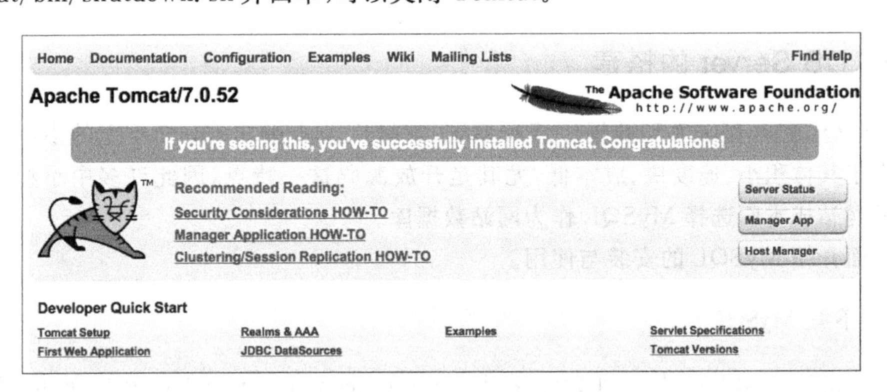

图 2.2 Tomcat 界面

2. 安装项目服务端文件

(1) 依次进入文件夹 Tomcat→webapps，如图 2.3 所示。

(2) 将本书电子资源中的 meServer 文件夹复制到 webapps 文件夹中，如图 2.4 所示。

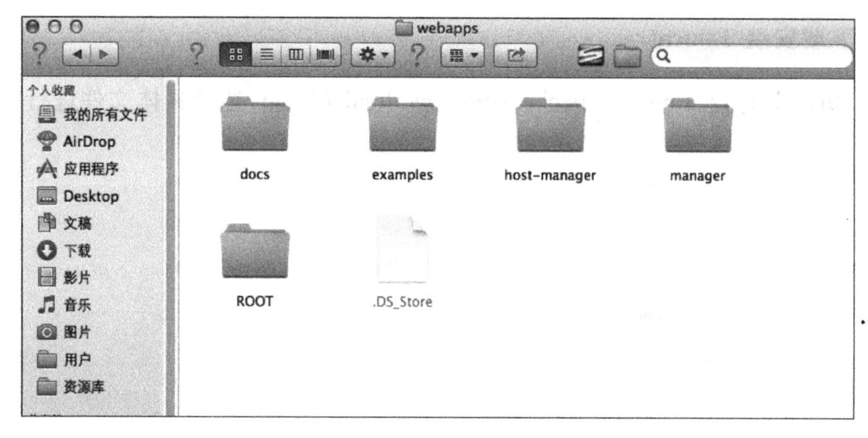

图 2.3 webapps 文件夹

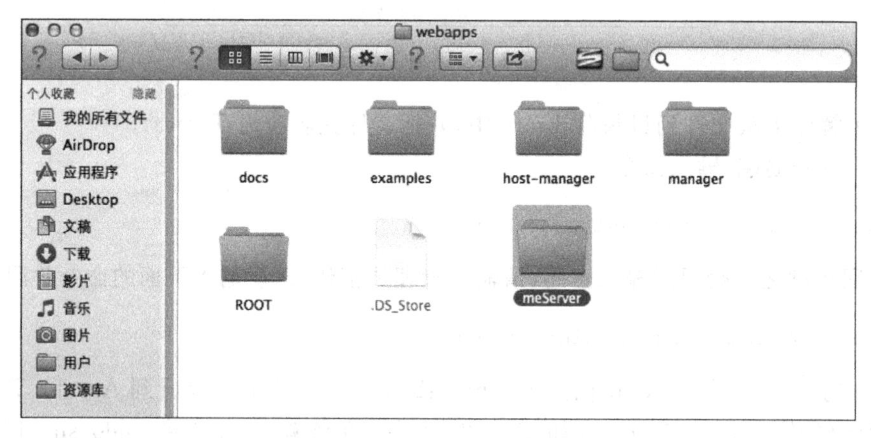

图 2.4 复制 meServer 文件夹到 webapps 文件夹

2.2.2 DB Server 的搭建

MySQL 是小型关系型数据库管理系统，它被广泛应用在 Internet 上的中小型网站中。由于其体积小、速度快、成本低，尤其是开放源码这一特点，因此许多中小型网站为了降低网站成本而选择 MySQL 作为网站数据库。

下面介绍 MySQL 的安装与使用。

1. 下载 MySQL

(1) 访问 MySQL 的官网 http://dev.mysql.com/downloads/，然后单击 MySQL Community Server 下方的 DOWNLOAD 选项，如图 2.5 所示。

(2) 在 Mac OS 上 MySQL 的版本很多，这里选择.dmg。单击右侧的 Download 按钮，如图 2.6 所示。

(3) 选择 No thanks,just start my download 直接下载，如图 2.7 所示。

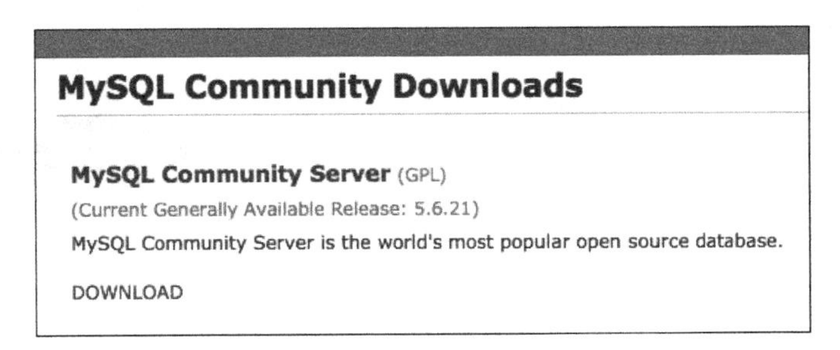

图 2.5 MySQL 下载页面

Mac OS X 10.9 (x86, 64-bit), Compressed TAR Archive	5.6.21	165.7M	Download
(mysql-5.6.21-osx10.9-x86_64.tar.gz)	MD5: 40b34ff61d8504b955ec4f016a8fe667 | Signature		
Mac OS X 10.9 (x86, 64-bit), DMG Archive	5.6.21	164.5M	Download
(mysql-5.6.21-osx10.9-x86_64.dmg)	MD5: 6f6b954c52cac4267a172c425f24fd8c | Signature		
Mac OS X 10.8 (x86, 64-bit), Compressed TAR Archive	5.6.21	165.7M	Download
(mysql-5.6.21-osx10.8-x86_64.tar.gz)	MD5: 40b34ff61d8504b955ec4f016a8fe667 | Signature		
Mac OS X 10.8 (x86, 64-bit), DMG Archive	5.6.21	164.5M	Download
(mysql-5.6.21-osx10.8-x86_64.dmg)	MD5: 6f6b954c52cac4267a172c425f24fd8c | Signature		

图 2.6 单击下载

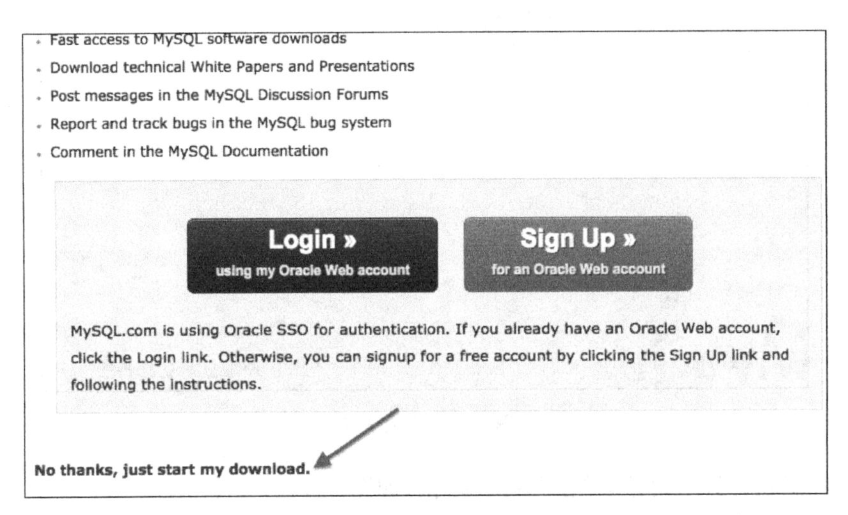

图 2.7 直接下载

2. 安装 MySQL

MySQL 的安装包如图 2.8 所示，安装步骤如下。

(1) 双击 mysql-5.6.21-osx10.8-x86_64.pkg 进行安装，单击“继续”按钮，如图 2.9 所示。

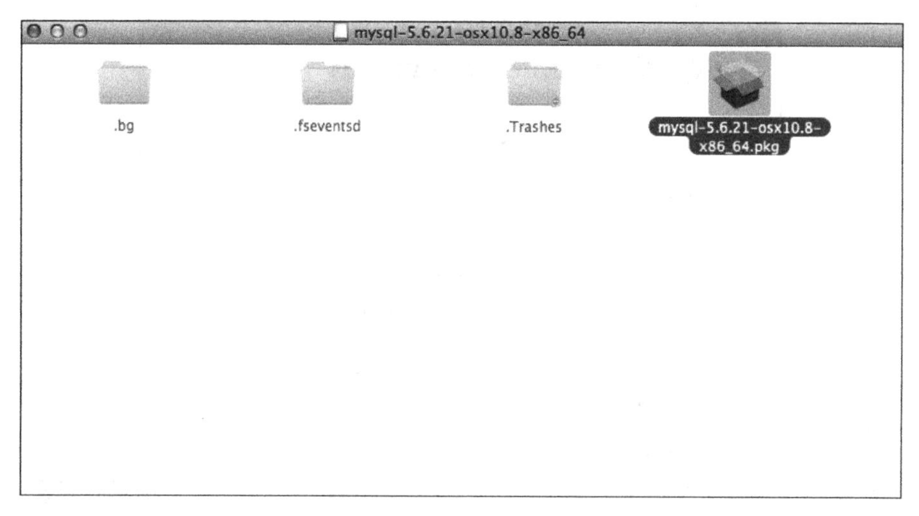

图 2.8　MySQL 安装包

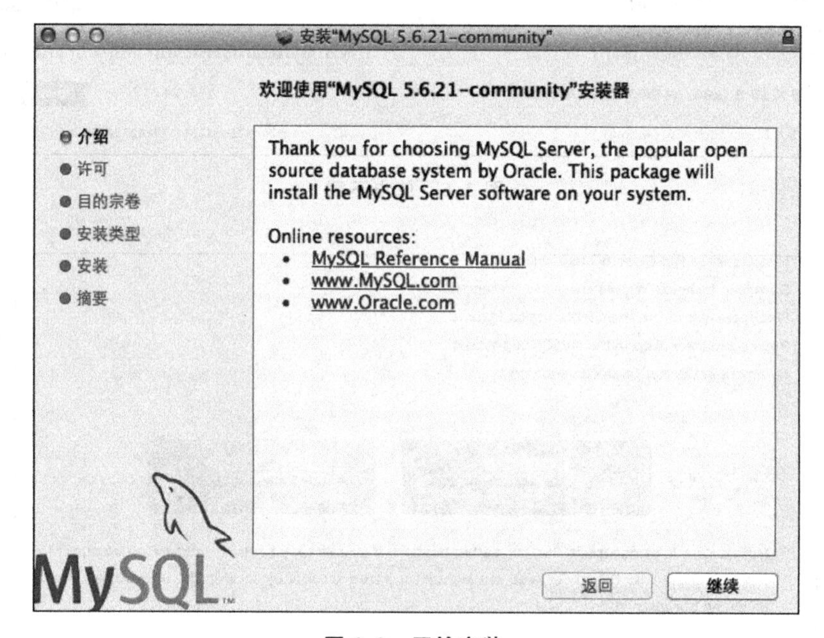

图 2.9　开始安装

(2) 同意软件许可协议，单击"继续"按钮，如图 2.10 所示。

(3) 选择安装类型，这里选择默认安装方式，单击"安装"按钮，如图 2.11 所示。

(4) 输入密码，单击"安装软件"按钮，如图 2.12 所示。

(5) 提示安装成功，单击"关闭"按钮，如图 2.13 所示。

3. 下载安装 MySQL Workbench(GUI Tool)

(1) 访问 http://dev.mysql.com/downloads/，找到 MySQL Workbench(GUI Tool)项，单击 DOWNLOAD 选项，进入下载界面，如图 2.14 所示。

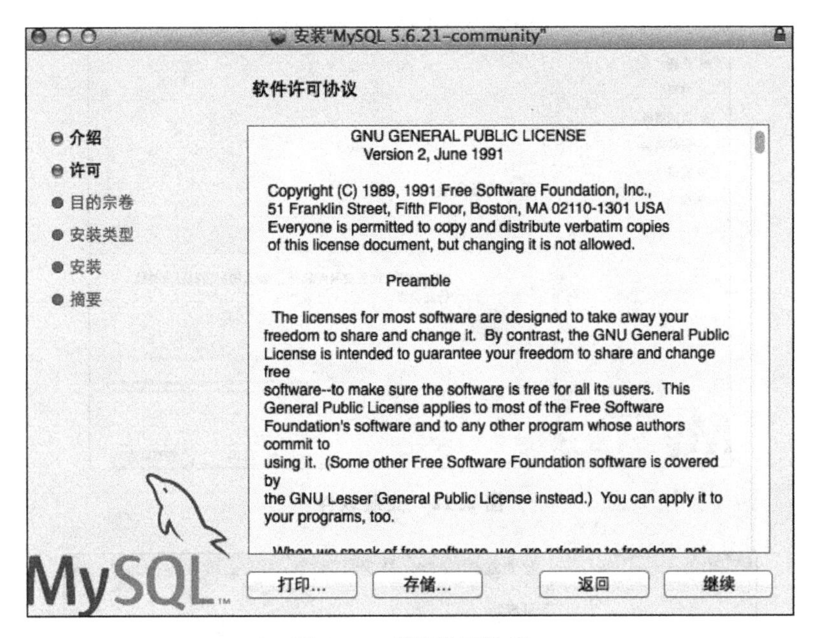

图 2.10　同意许可协议

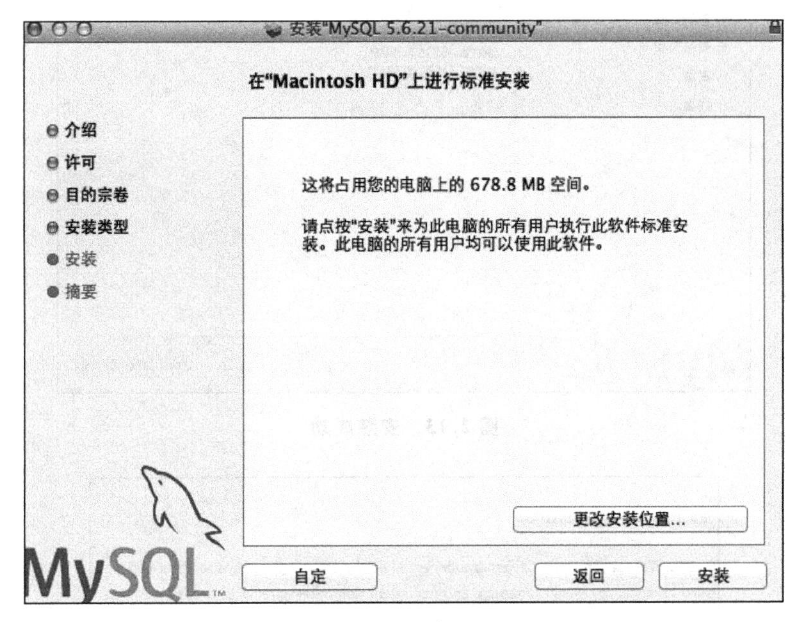

图 2.11　选择安装类型

图 2.12　安装软件

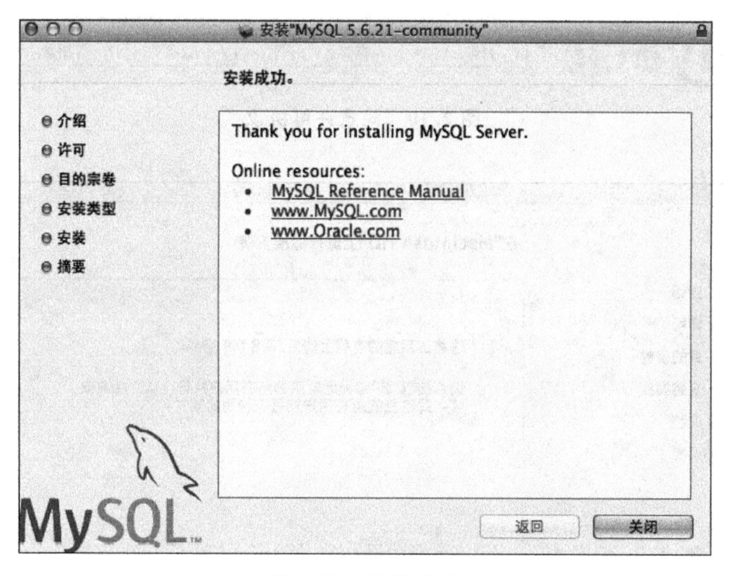

图 2.13　安装成功

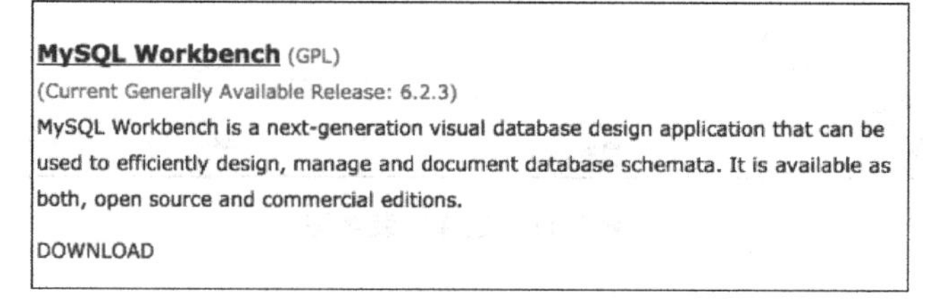

图 2.14　MySQL Workbench

(2) 选择版本之后单击 Download 按钮，如图 2.15 所示。

(3) 选择 No thanks,just start my download 直接下载，如图 2.16 所示。

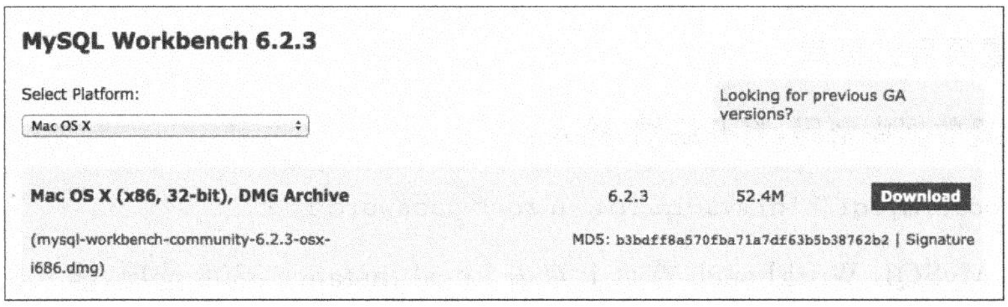

MySQL Workbench 6.2.3

Select Platform:

Mac OS X

Looking for previous GA versions?

Mac OS X (x86, 32-bit), DMG Archive 6.2.3 52.4M Download

(mysql-workbench-community-6.2.3-osx-i686.dmg) MD5: b3bdff8a570fba71a7df63b5b38762b2 | Signature

图 2.15 单击下载

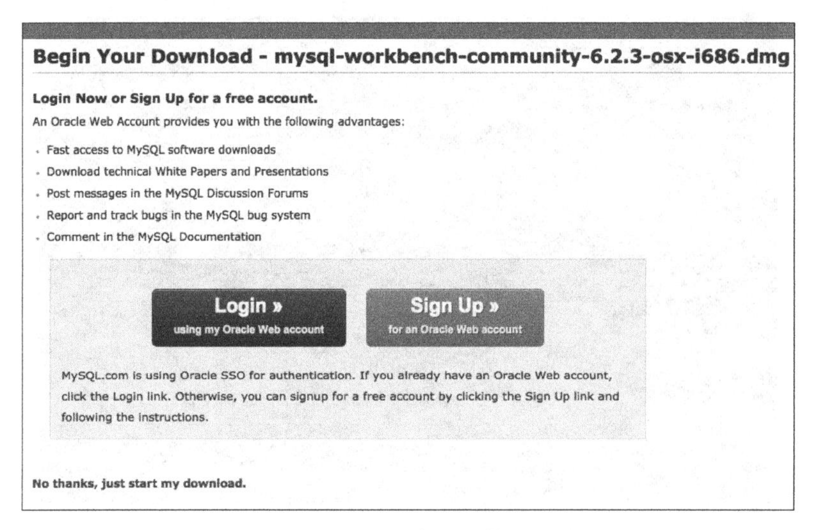

图 2.16 直接下载

(4) 下载完成之后双击即可安装。安装完成之后在“应用程序”里面就能看到 MySQL Workbench.app 程序，双击打开，如图 2.17 所示。

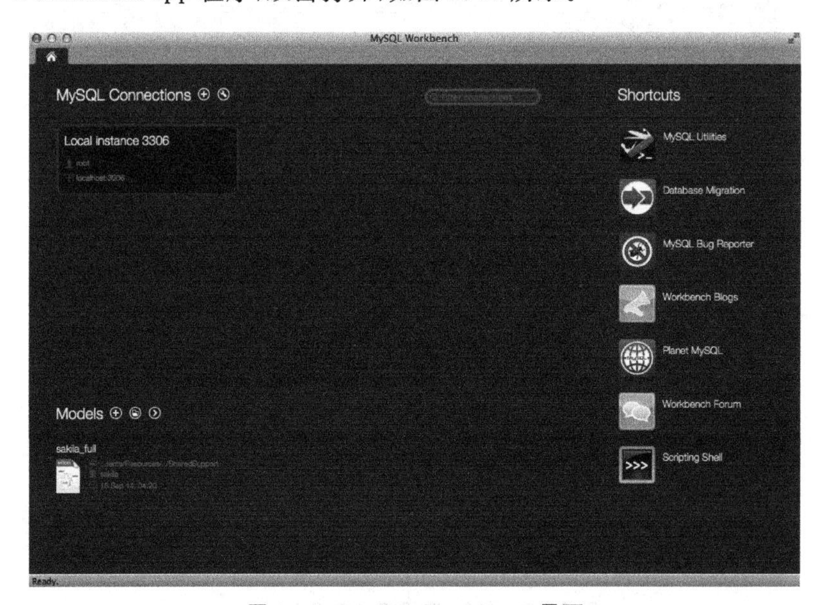

图 2.17 MySQL Workbench 界面

4. 配置项目数据库

(1) 设置数据库密码,打开终端,输入:

```
/usr/local/mysql/bin/mysqladmin-u root password root
```

(2) 在 MySQL Workbench 首页上双击 Local instance 3306 选项,输入密码 root,如图 2.18 所示。

图 2.18　双击 Local instance 3306

(3) 在 SCHEMAS 下方右击,选择 Create Schema...,如图 2.19 所示。

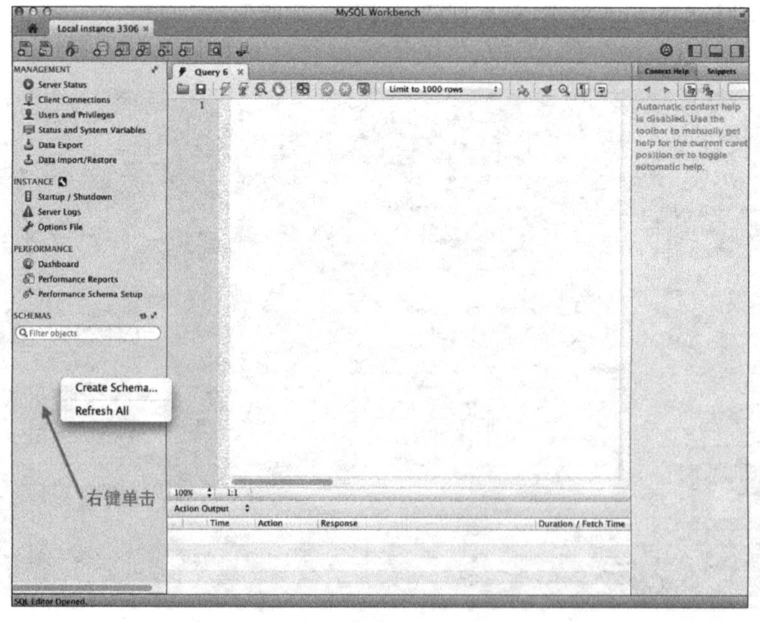

图 2.19　右击,选择 Create Schema...

(4) 在 Schema Name 文本框中输入 me_server，单击 Apply 按钮，如图 2.20 所示。

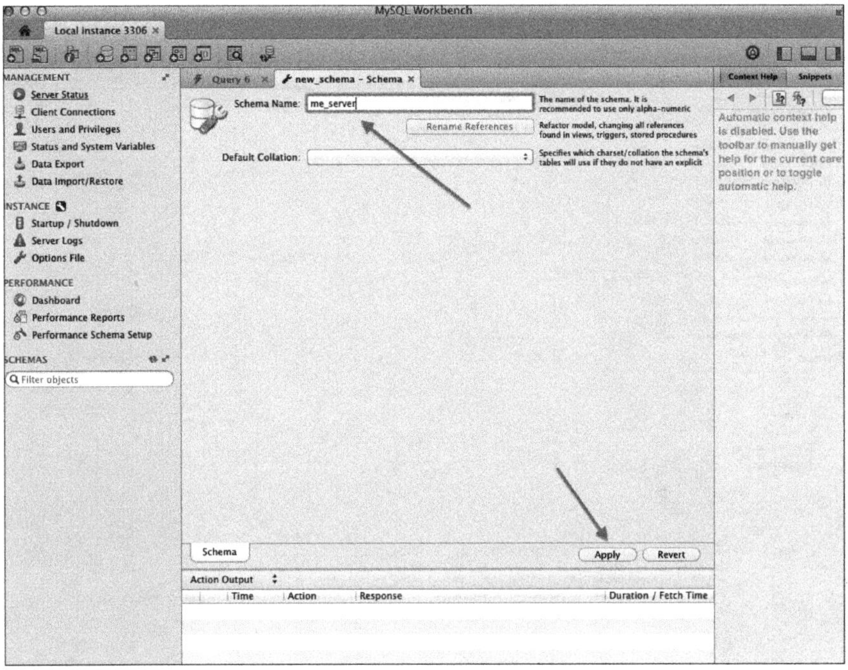

图 2.20 输入 Schema Name

(5) 继续单击 Apply 按钮，如图 2.21 所示，然后单击 Close 按钮完成创建。

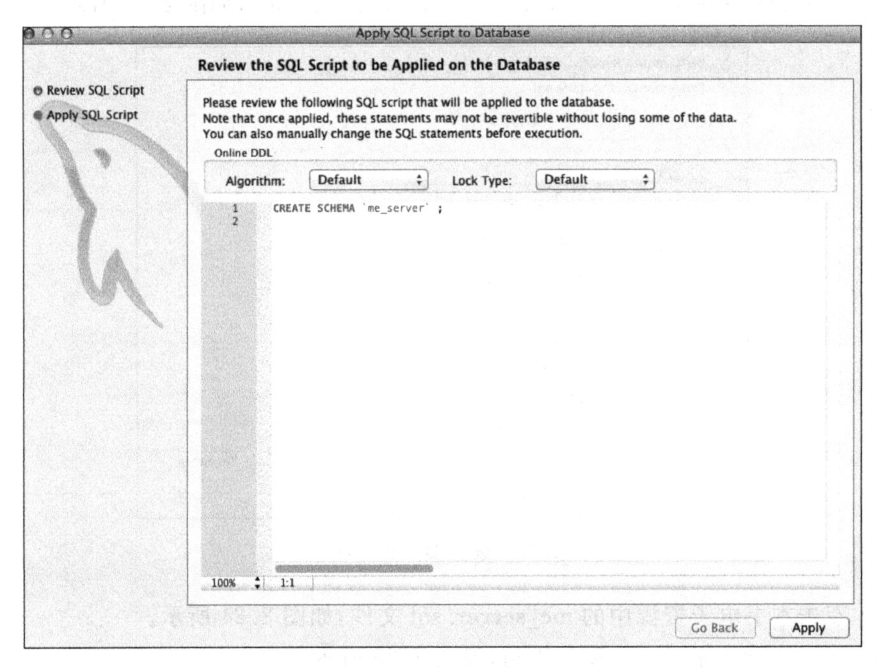

图 2.21 单击 Apply

(6) 创建成功后的界面如图 2.22 所示。

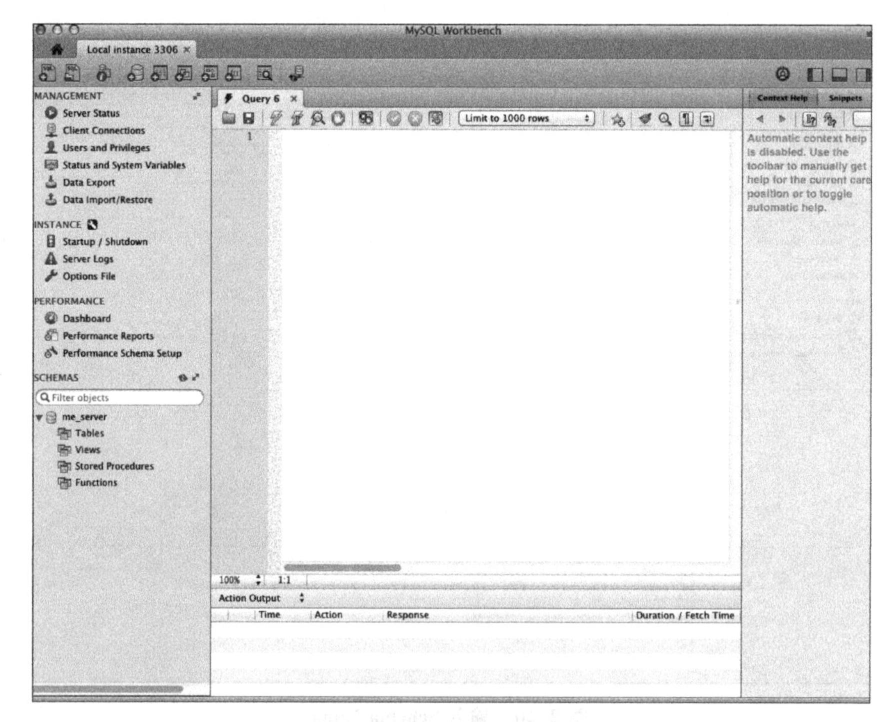

图 2.22　创建成功

(7) 右击 me_server 数据库,选择 Set as Default Schema,如图 2.23 所示。

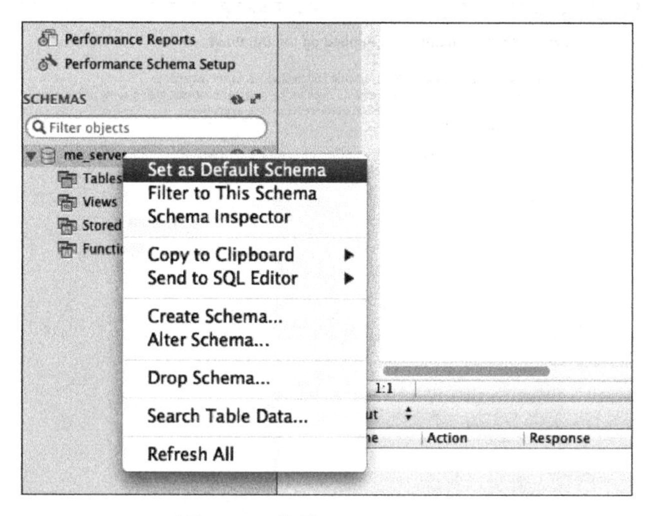

图 2.23　设置 Default Schema

(8) 双击本书电子资源中的 me_server.sql 文件,如图 2.24 所示。

(9) 单击执行按钮,执行 SQL 语句,如图 2.25 所示。

(10) 右击 me_server 数据库,选择 Refresh All 刷新数据,如图 2.26 所示。

(11) 数据库下的表创建成功,如图 2.27 所示。

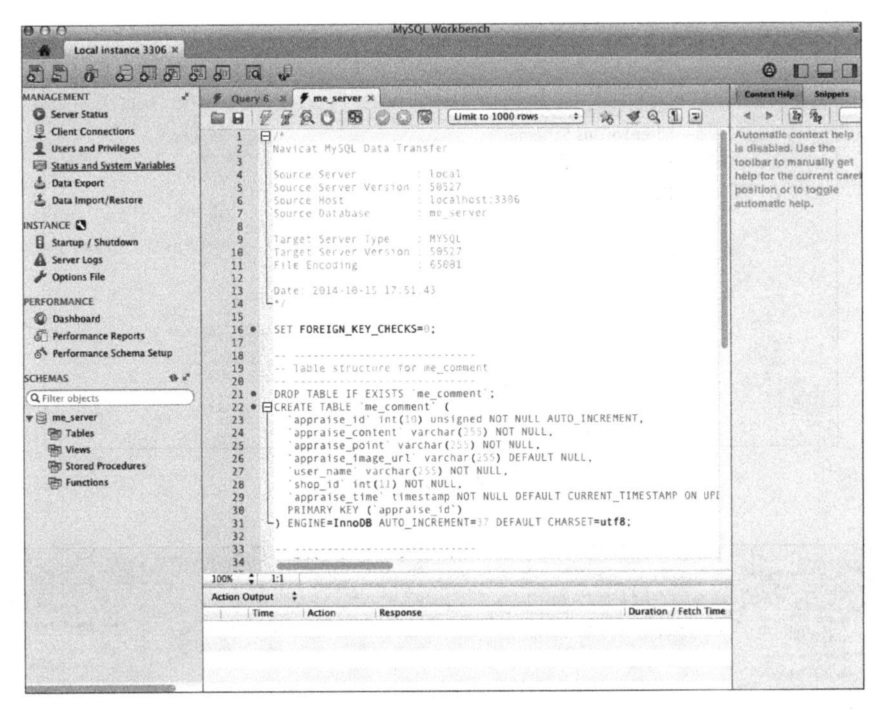

图 2.24 me_server.sql 文件

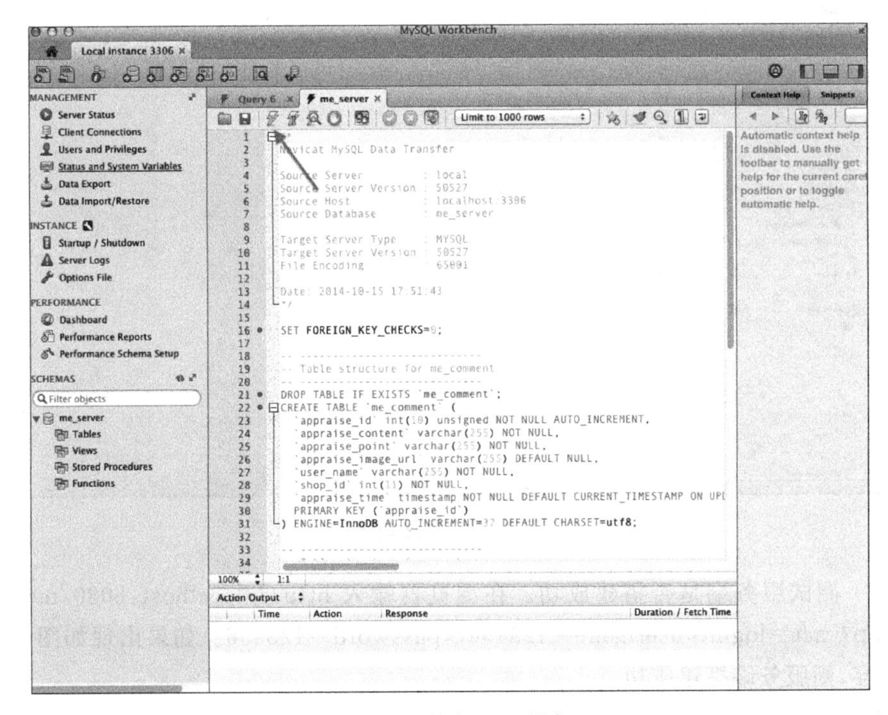

图 2.25 执行 SQL 语句

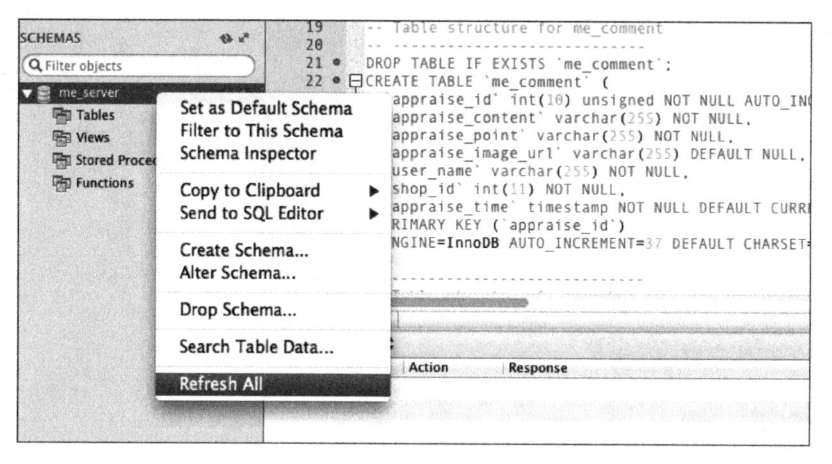

图 2.26　刷新数据

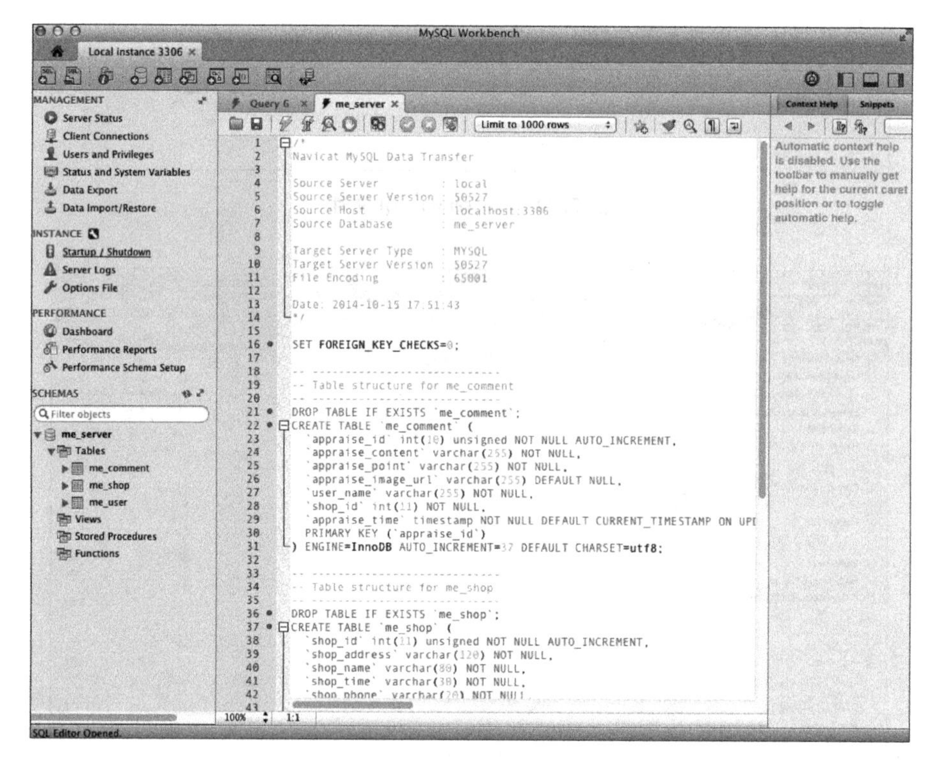

图 2.27　最终效果

(12) 测试服务端是否搭建成功。在浏览器输入 http://localhost:8080/meServer/user.php? act=login&username=123456&password=123456。如果出现如图 2.28 所示的内容，则服务端搭建成功。

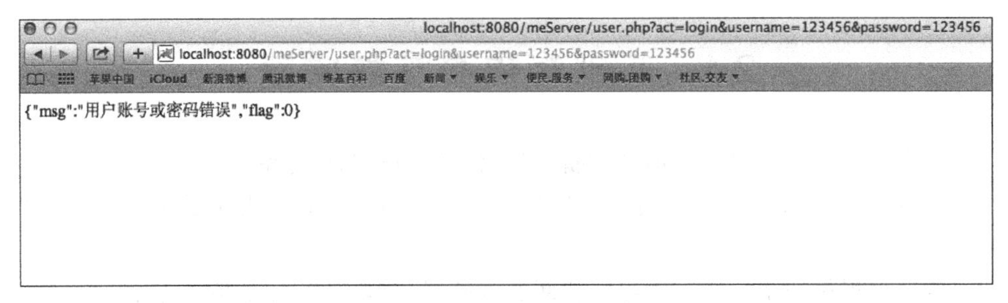

图 2.28 测试服务端是否搭建成功

2.3 几个必备的基础技能

(1) 对编程语言有所了解,例如 C/C++ 。

(2) 掌握 iOS 开发语言 Objective-C。

Objective-C 是非常实际的语言。它是一个用 C 写成、很小的运行库,不会过分地增加应用程序的大小,与大部分面向对象的系统使用较长的 VM 执行时间会取代整个系统的运作相反,Objective-C 写成的程序通常不会比其原始代码大很多。

(3) 熟悉面向对象思想。面向对象是一种对现实世界理解和抽象的方法,是计算机编程技术发展到一定阶段的产物。早期的计算机编程是基于面向过程的方法,例如实现算术运算 1+1+2=4,通过设计一个算法就可以解决当时的问题。随着计算机技术的不断发展,计算机被用于解决越来越复杂的问题。一切事物皆对象,通过面向对象的方式,将现实世界的事物抽象成对象,现实世界中的关系抽象成类、继承,可帮助人们实现对现实世界的抽象与数字建模。通过面向对象的方法,更利于用人理解的方式对复杂系统进行分析、设计与编程。同时,面向对象能有效提高编程的效率,通过封装技术,消息机制可以像搭积木一样快速开发出一个全新的系统。面向对象是指一种程序设计范型,同时也是一种程序开发的方法。对象指的是类的集合。它将对象作为程序的基本单元,将程序和数据封装其中,以提高软件的重用性、灵活性和扩展性。

起初,面向对象是专指在程序设计中采用封装、继承、多态等设计方法。

面向对象的思想已经涉及软件开发的各个方面,如面向对象的分析(Object Oriented Analysis,OOA)、面向对象的设计(Object Oriented Design,OOD)以及面向对象的编程实现(Object Oriented Programming,OOP)。

面向对象的分析根据抽象关键的问题域来分解系统。面向对象的设计是一种提供符号设计系统的面向对象的实现过程,它用非常接近实际领域术语的方法把系统构造成"现实世界"的对象。面向对象程序设计可以看作一种在程序中包含各种独立而又互相调用的对象的思想,这与传统的思想刚好相反。传统的程序设计主张将程序看作一系列函数的集合,或者直接就是一系列对计算机下达的指令。面向对象程序设计中的每一个对象都应该能够接受数据、处理数据并将数据传送给其他对象,因此它们都可以被看作一个小型的"机器",即对象。

2.4 基础知识与技能回顾

本章介绍了iOS项目开发前的准备工作，包括客户端、服务端。

服务端采用Tomcat+MySQL的方式，用来模拟真实项目开发中手机客户端与服务端的数据交互。

一定要确保按照本章所讲内容成功部署服务端，否则项目无法继续进行。

练　习

项目

功能描述：搭建本地服务端环境并测试。

第3章 chapter 3

MyDemo项目介绍

本章从项目背景、项目需求分析、项目用例分析、项目数据库来介绍项目前期的准备工作，以便对本项目有一个全局的认识，这也是项目开发的一般流程。

3.1 项目背景

中国互联网络信息中心发布的第41次《中国互联网络发展状况统计报告》指出，截至2017年12月，我国网民规模达7.72亿，互联网普及率为55.8%，较2016年底提升2.6个百分点。我国手机网民规模达7.53亿，较2016年底增加5734万人。网民中使用手机上网人群的占比由2016年的95.1%提升至97.5%。

以上数据明确表明：移动互联网的发展已成为不可阻挡的趋势。

3.2 项目需求分析

在这个互联网高速发展的时代，越来越多电子商务业务的出现，大大提高了生活的便利性。B2B、B2C、C2C等各种业务形态如雨后春笋般地出现，移动互联网作为未来发展业务、拓展业务的利器，是真正实现物联网的关键部分。

本项目是一款基于LBS的类团购应用，主要包含商户展示、商户搜索、商户浏览、用户评价等功能，后台能对商户信息进行增加、删除、更新等操作。用户可以实现查看商户最新的促销信息，搜索商户，评价商户等功能。

3.3 项目用例分析

用例分析是从用例模型到分析模型的过程，是需求与设计之间的桥梁。用例分析把系统的行为分配给分析类，让分析类交互完成系统的行为。

本项目的用例分析分为用户和后台两部分，如图3.1、图3.2所示。

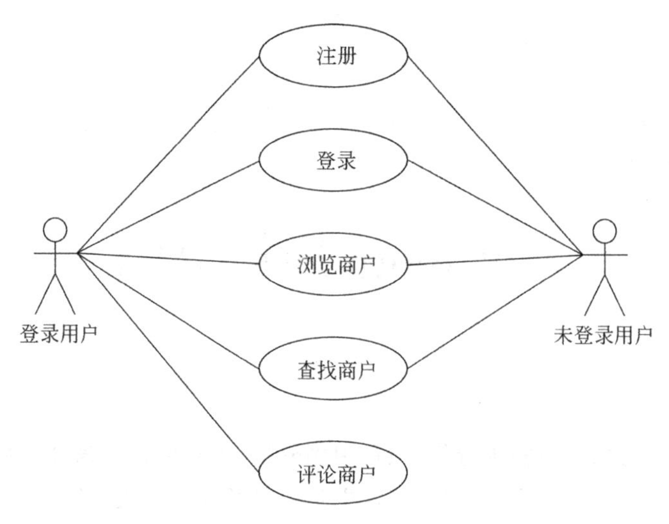

图 3.1 用户用例分析

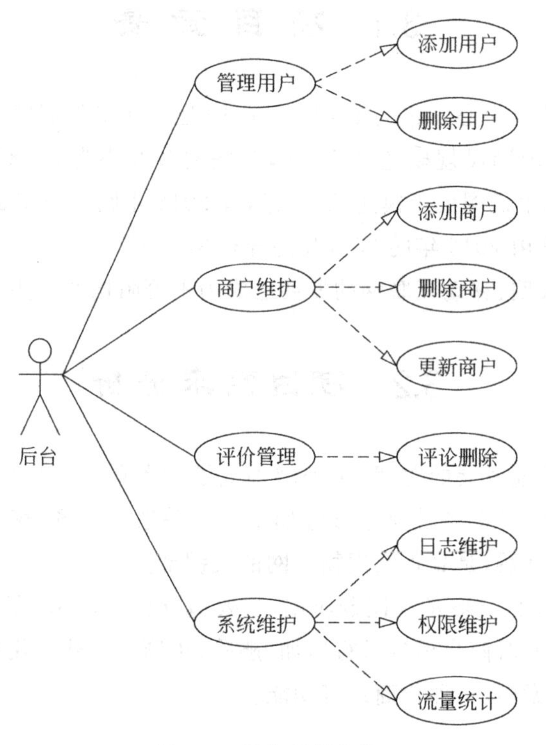

图 3.2 后台用例分析

3.4 项目数据库分析

在开发数据库的时候，首先要做的就是识别实体以及实体之间的关系，并将实体间的联系在数据库表中用表及主外键约束表示出来。E-R 图的作用就是为了更有效地在概念模式下设计数据库，更形象地表示实体及实体之间的关系。

本项目的数据库 E-R 图如图 3.3 所示。

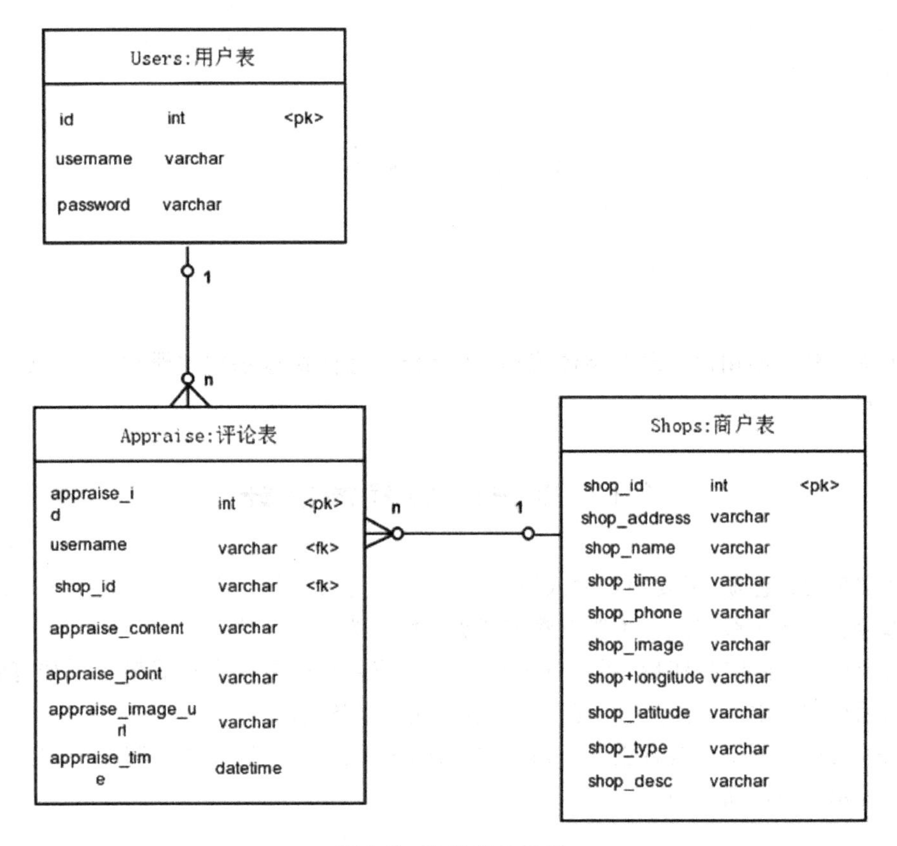

图 3.3　数据库 E-R 图

3.5　基础知识与技能回顾

本章介绍项目开发前的准备工作，包括项目背景、项目需求分析、项目用例分析、项目数据库分析。良好的分析活动有助于避免或尽早剔除早期错误，从而提高软件生产率，降低开发成本，改进软件质量。

练　　习

项目

功能描述：为团购类应用写一份简单的需求分析。

第4章 chapter 4

用户注册

本章首先介绍用户注册的整体流程，然后介绍怎样快速添加注册功能，并实现与服务端交互。

4.1 用户注册总体设计

要实现用户注册，需要用户输入用户名、密码和重复密码。其中用户名为6～12位，由字母、数字、下画线组成且不能以数字开头。密码为6～25位。

用户单击“注册”按钮后，系统首先检查用户信息是否按规定格式输入。如有错误，则提示用户重新输入。如正确，则将输入信息提交服务端。

服务端检查用户名是否重复。如果重复，则提示注册失败；反之，将用户信息写入数据库并提示用户注册成功。

4.1.1 流程图

用户注册流程如图4.1所示。

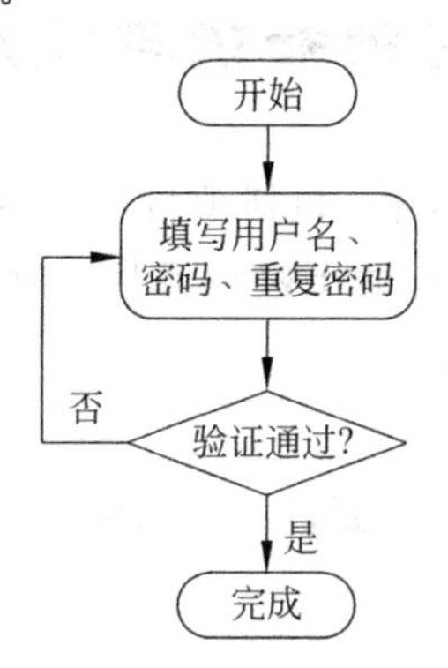

图4.1 用户注册流程

4.1.2 时序图

用户注册的时序过程如图4.2所示。

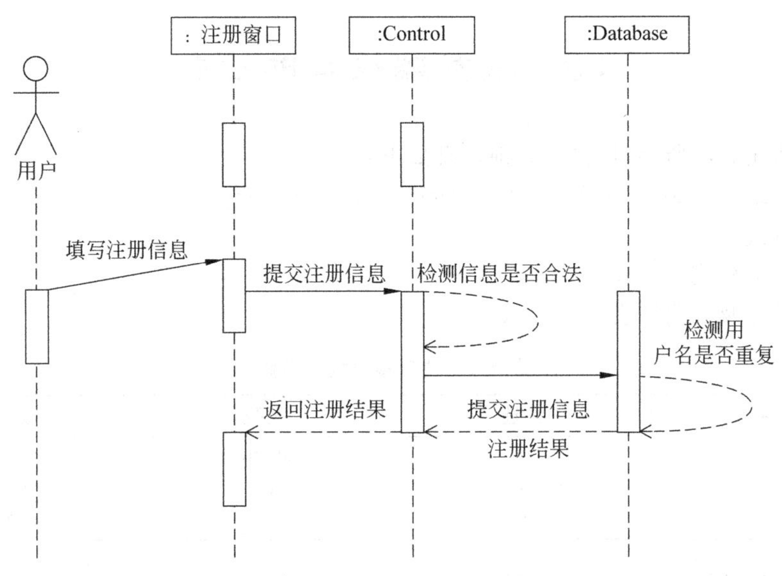

图 4.2 注册时序过程

4.2 数据库的准备

第 2 章已经介绍了如何搭建项目服务端并配置完成本项目所需后台数据。下面简要介绍数据库的准备。

用户表用来存储用户信息,如图 4.3 所示。

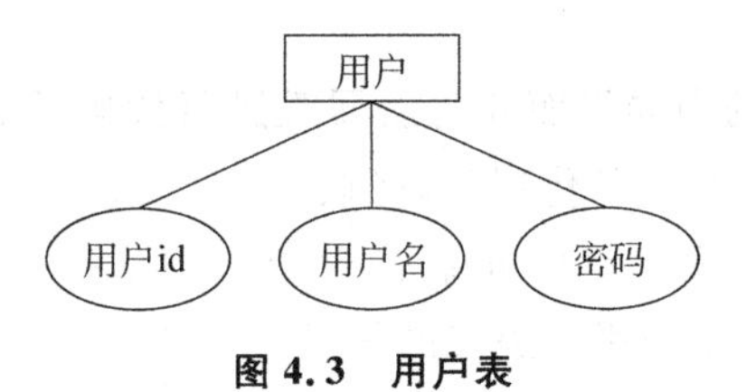

图 4.3 用户表

SQL 语句如下:

```
CREATE TABLE 'me_user' (
    'id' int(11) unsigned NOT NULL AUTO_INCREMENT,
    'username' varchar(255) DEFAULT NULL,
    'password' varchar(255) DEFAULT NULL,
    PRIMARY KEY ('id')
) ENGINE= InnoDB AUTO_INCREMENT= 4 DEFAULT CHARSET= utf8;
```

4.3 服务端接口的准备

本节将用到注册接口，接口详细信息如下：
接口地址：http://localhost:8080/meServer/user.php?
调用方式：Post
返回数据格式：Json
提交参数及说明如表4.1所示。

表4.1 注册接口提交参数

请求参数	必选	类型	说明
act	Y	string	register
username	Y	string	用户名
password	Y	string	密码

返回字段及说明如表4.2所示。

表4.2 注册接口返回字段

返回字段	字段类型	字段说明
flag	string	0：失败，1：成功
msg	string	信息说明

4.4 用户注册的实现

本节首先介绍用xib的布局界面并与所属类进行代码关联。与纯代码的布局相比，xib布局的最大好处是大量减少了代码量，降低了布局难度，方便初学者快速开发。然后介绍客户端与服务端的交互，让服务端来验证用户提交的信息。

用户注册界面如图4.4所示。

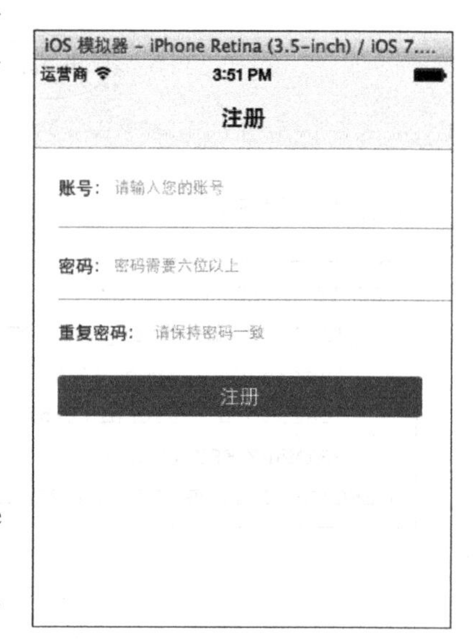

图4.4 用户注册界面

4.4.1 客户端代码开发

一个iOS应用的所有文件都在一个Xcode项目下。下面创建一个新项目。

1. 新建工程

(1) 打开Xcode，单击Create a new Xcode project选项，如图4.5所示。

(2) 在Choose options for your new project对话框中依次输入项目名称、机构名。在Devices的

图 4.5 创建项目

下拉列表中选择 iPhone，不要勾选 Use Core Data，如图 4.6 所示。然后单击 Next 按钮。

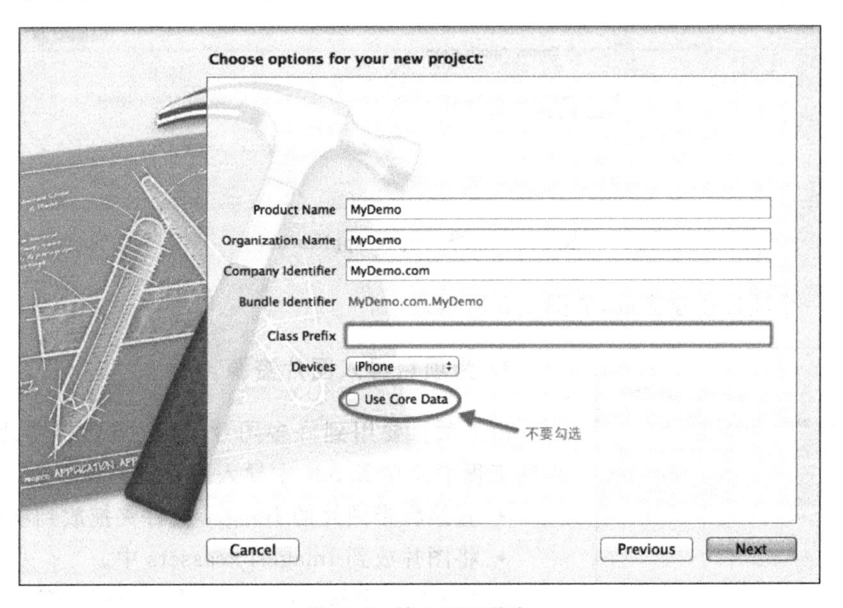

图 4.6 输入项目信息

在图 4.6 所示对话框的 Devices 下拉列表中，可以选择 iPhone 或者 iPad。iPhone 和 iPad 是不同的。iPad 有更大的屏幕，而且有一些 iPhone 上没有的东西，如分割视图和弹出视图。因此，一个 iPad 项目的 nib 文件和 iPhone 项目是不同的。可以为项目选择几个目标设备。

- iPad：应用程序只运行在 iPad 上。

- iPhone：应用程序可以运行在 iPhone 或者 iPod Touch 上，也能运行在 iPad 上（作为一个兼容模式）。
- Universal：应用程序在 iPhone 和 iPad 上都能运行，是一个通用的程序。对于一个通用的应用程序，在不同设备上会载入不同的资源文件。

（3）出现 Choose a template for your new project 对话框，如图 4.7 所示。此时，在左边选择 iOS 下面的 Application，然后右边出现了多个模板，如主从应用（Master-Detail Application）、基于页面的应用（Page-Based Application）、单视图应用（Single View Application）等。选择 Empty Application，单击 Next 按钮来创建一个空模板的应用。

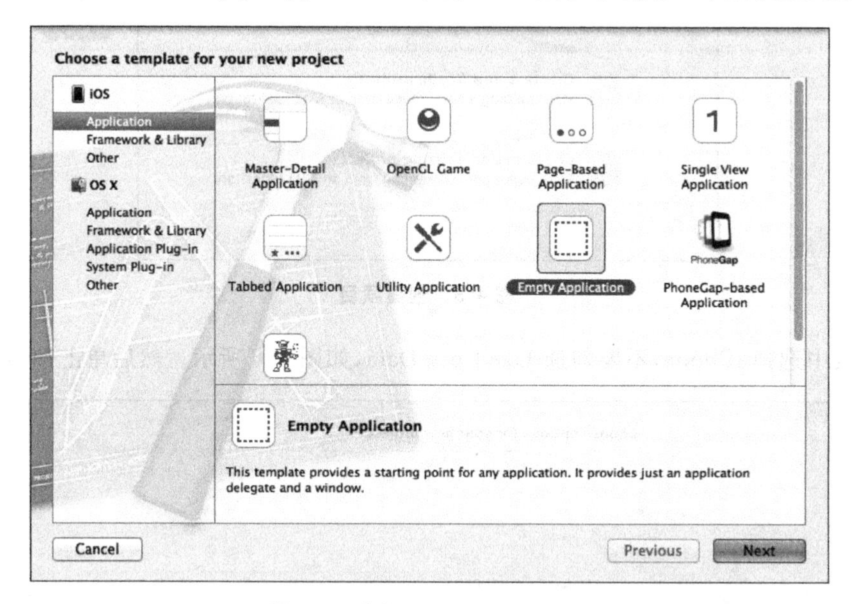

图 4.7 选择 Empty Application

（4）一个项目创建完成，如图 4.8 所示。

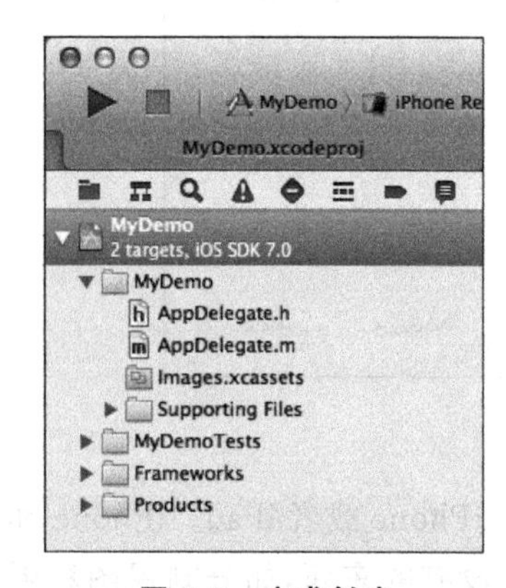

图 4.8 完成创建

2. 为项目导入图片资源

由于项目要用到许多图片，因此需要将图片资源加入到工程中。在 Xcode 中导入图片有两种方法：

- 直接将带图片的 Images 文件夹拖放到项目中。
- 将图片放到 Images. xcassets 中。

第 1 种方法操作简单。第 2 种方法便于管理和维护。如果项目最低支持 iOS7，则可以在 Image. xcassets 内制定图片的 slicing，也就是拉伸图片时涉及的 capInsets。

这里使用第 1 种方法，直接将图片文件夹导入到项目中。

（1）打开本书的电子资源，拖动 Images 文件夹到项目中，如图 4.9 所示。

（2）勾选 Copy items into destination group's folder(if needed)，如图 4.10 所示。

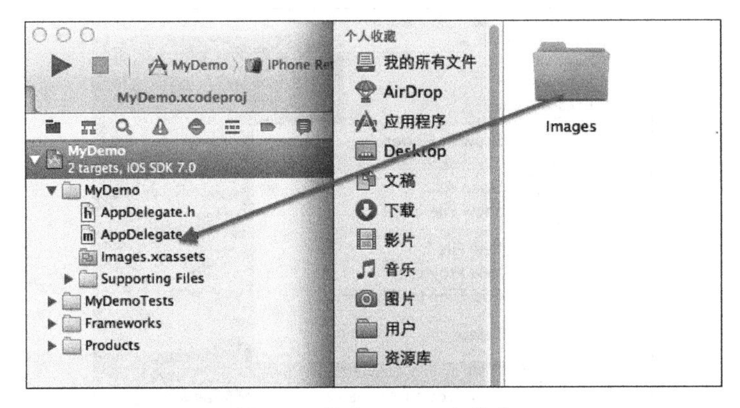

图 4.9 拖曳 Images 文件夹

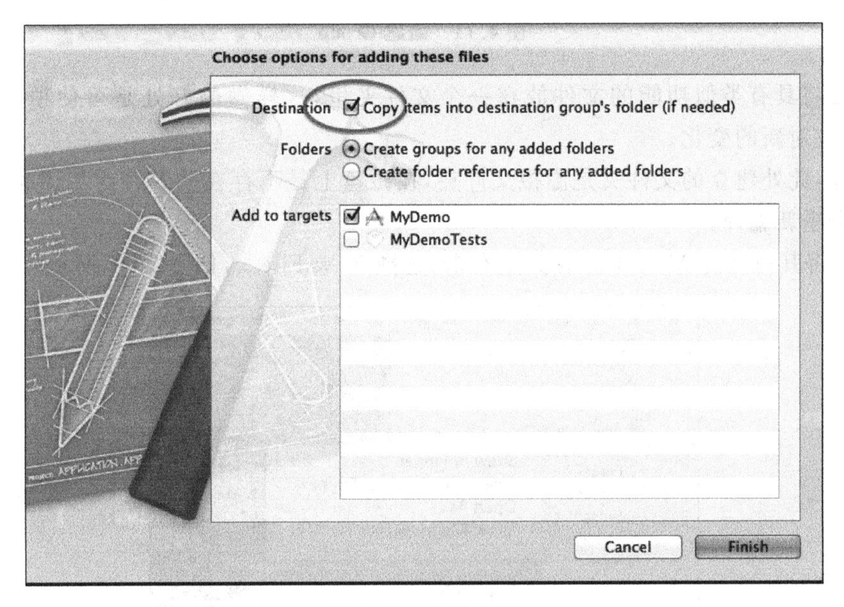

图 4.10 注意勾选

Copy items into destination group’s folder(if needed)：将 Images 文件夹复制到工程物理文件夹下。由于 Images 文件夹不在工程文件夹中，所以此处要勾选。

Create groups for any added folders：把选择的文件添加到工程的 group 下。如果选择的是文件，则把文件夹看作 group。这个文件夹对应的工程目录和文件路径不一定是一一对应的。文件夹的颜色为黄色。

Create folder references for any added folders：建立一个文件夹的索引，同时文件夹中的所有文件也会添加到整个工程中。这个文件夹的工程目录和文件路径是一一对应的。文件夹的颜色为红色。

3. 为项目添加宏定义

(1) 右击 AppDelegate. m，选择 New Group，如图 4.11 所示。

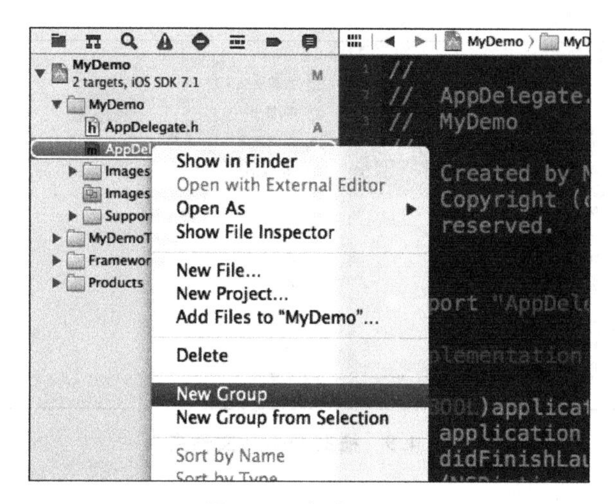

图 4.11 新建 Group

通常将具有类似功能的文件放在一个文件夹里，这样做的好处是可使项目结构清晰，容易应对新的变化。

另外，此处建立的文件夹是虚拟文件夹，在磁盘上是不存在的。每次编译时，这类文件夹都会重新编译。

(2) 将其命名为 Macro，并在其上右击，选择 New File...，如图 4.12 所示。

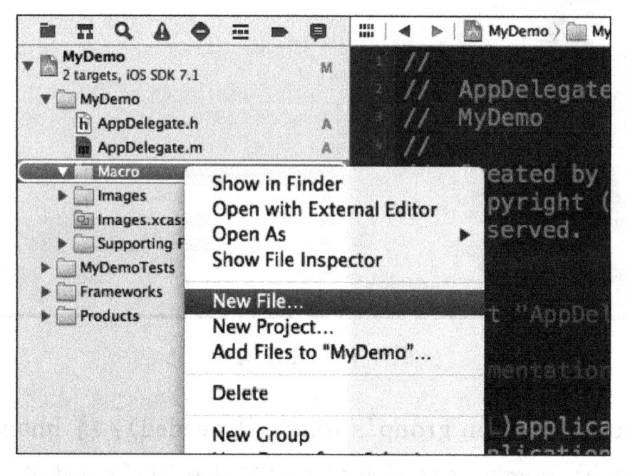

图 4.12 新建 File

Macro 文件夹用来存放宏定义文件。什么是宏定义呢？宏定义又称为宏代换、宏替换，简称“宏”。例如：

```
#define 标识符 字符串
```

其中的标识符就是符号常量，也称为“宏名”。

iOS 开发过程中使用宏可以提高开发效率，提高代码的重用性。比如在开发中，经常需要获取屏幕高度，一般写法：

```
float height=[UIScreen mainScreen].bounds.size.height;
```

使用宏定义：

```
#define App_SCREEN_HEIGHT [UIScreen mainScreen].bounds.size.height
```

只需这样写：

```
float height=App_SCREEN_HEIGHT
```

(3) 选择 C and C++ 下的 Header File，单击 Next 按钮，如图 4.13 所示。

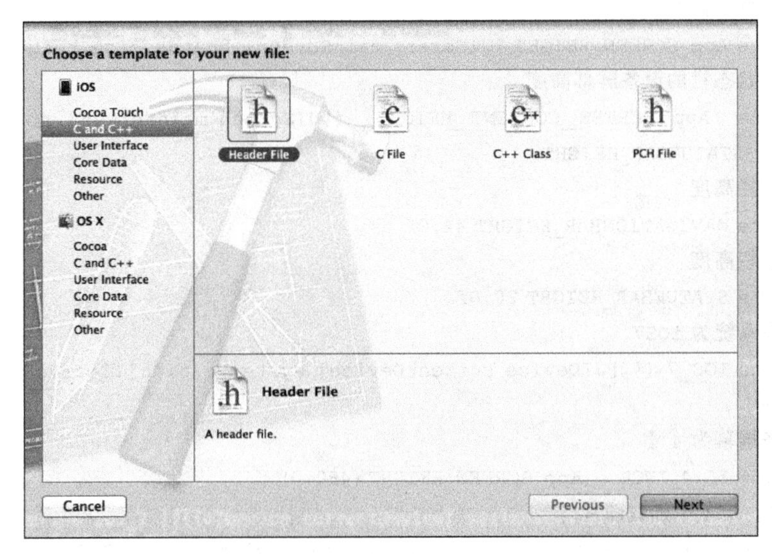

图 4.13 创建 Header File

Header File 是一个头文件，这里将宏定义写到这个头文件中，然后在用到头文件的地方就可以引用这个头文件。

(4) 在 Save As 文本框中输入 UtilsMacro.h，在 Targets 中勾选 MyDemo，如图 4.14 所示。

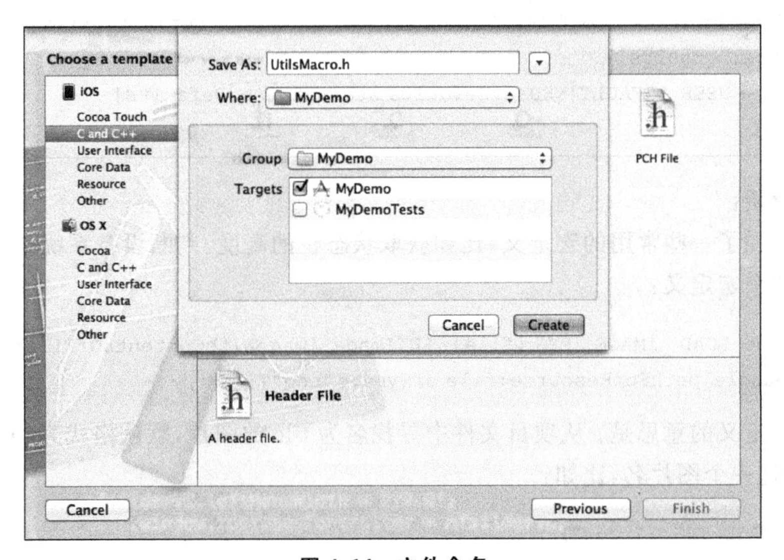

图 4.14 文件命名

(5) 单击 UtilsMacro.h,修改代码如下：

```
#ifndef MyDemo_UtilsMacro_h
#define MyDemo_UtilsMacro_h

//设备屏幕宽度
#define App_SCREEN_WIDTH [UIScreen mainScreen].bounds.size.width
//设备屏幕高度
#define App_SCREEN_HEIGHT [UIScreen mainScreen].bounds.size.height
//除去状态栏的设备屏幕高度
#define App_SCREEN_CONTENT_HEIGHT ([UIScreen mainScreen].bounds.size.
height-STATUEBAR_HEIGHT)
//导航栏高度
#define NAVIGATIONBAR_HEIGHT 44.0f
//状态栏高度
#define STATUEBAR_HEIGHT 20.0f
//设备系统为 iOS7
#define IOS_7 ([[[UIDevice currentDevice] systemVersion]floatValue]>=7.0?
YES:NO)
//设备屏幕为 4 寸
#define IS_4_INCH (App_SCREEN_HEIGHT>480.0)
//从 Bundle 中加载图片
#define LOAD_IMAGE(file,type) [UIImage imageWithContentsOfFile:[[NSBundle
mainBundle]pathForResource:file ofType:type]]
//从 Bundle 中加载.png 格式的图片
#define LOAD_IMAGE_PNG(file) [UIImage imageWithContentsOfFile:[[NSBundle
mainBundle]pathForResource:file ofType:@"png"]]
//通知
#define NOTIFICATION_CENTER [NSNotificationCenter defaultCenter]
//NSUserDefaults
#define USER_DEFAULT[NSUserDefaults standardUserDefaults]
#endif
```

代码解析：

这里添加了一些常用的宏定义,比如获取状态栏的高度,判断设备系统是否为 iOS7。注意下面这个宏定义：

```
#define LOAD_IMAGE_PNG(file) [UIImage imageWithContentsOfFile:[[NSBundle
mainBundle]pathForResource:file ofType:@"png"]]
```

这个宏定义的意思是：从项目文件中寻找名为 file 的图片,默认格式为 png。在使用中,需要传递一个图片名,比如：

```
LOAD_IMAGE_PNG(@"icon")
```

(6) 单击 Supporting Files 文件夹下的 MyDemo-Prefix. pch，导入头文件 UtilsMacro. h，如图 4.15 所示。

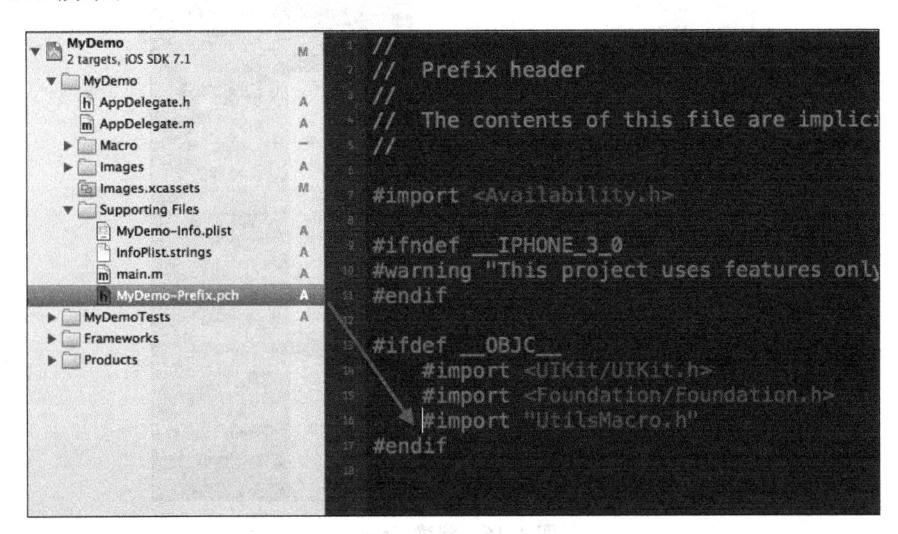

图 4.15 导入文件

新建一个工程时，在 Supporting Files 文件下会看到一个以-Prefix. pch 结尾的文件。pch 的全称是 precompiled header，也就是预编译头文件。该文件存放工程中一些不需经常修改的代码，比如常用的框架头文件。这样做的目的是提高编译器的编译速度。

修改工程中某个文件代码时，编译器不是重新编译所有文件，而是编译改动过的文件。假如 pch 中某个文件修改了，那么 pch 整个文件里包含的其他文件也会重新编译一次，这样就会消耗大量的时间，所以它里面添加的文件是很少变动或不变动的头文件或者是预编译的代码片段。

另外，在这里导入 UtilsMacro. h 头文件后，整个工程中都可以使用。

4. 为项目添加 Category

无论一个类的设计如何完美，都不可避免地会遇到没有预测到的需求。怎么扩展现有的类呢？当然，继承是个不错的选择，但 Objective-C 提供了一种特别的方式来扩展类，叫做 Category。它可以动态地为已经存在的类添加新的行为。这样可以保证类在原来的基础上做较小的改动就可以增加需要的功能。

使用 Category 对类进行扩展时，不需要访问其源代码，也不需要创建子类，这样就可以扩展系统提供的类。Category 使用简单的方式，实现了类的相关方法的模块化，从而把不同的类方法分配到不同的分类文件中。

下面介绍如何创建一个 UIView 的 Category，方便地获取 UIView 的各种坐标属性。

(1) 右击 AppDelegate. m，选择 New Group，如图 4.16 所示。

(2) 将其命名为 Categories，并在其上右击，选择 New File...，如图 4.17 所示。

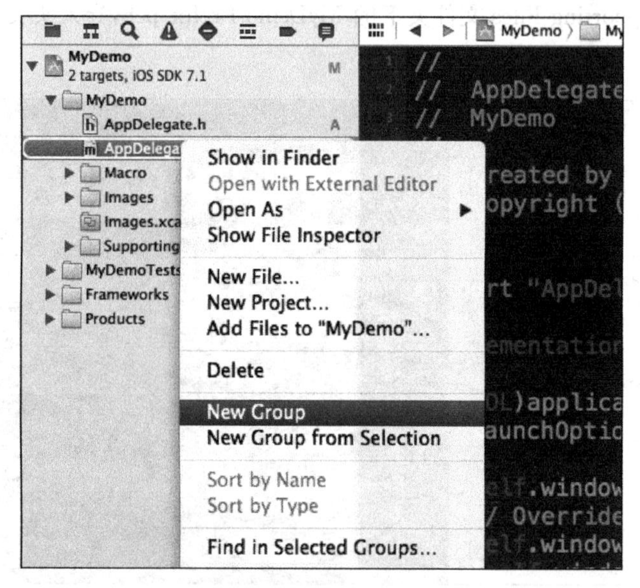

图 4.16 创建 Group

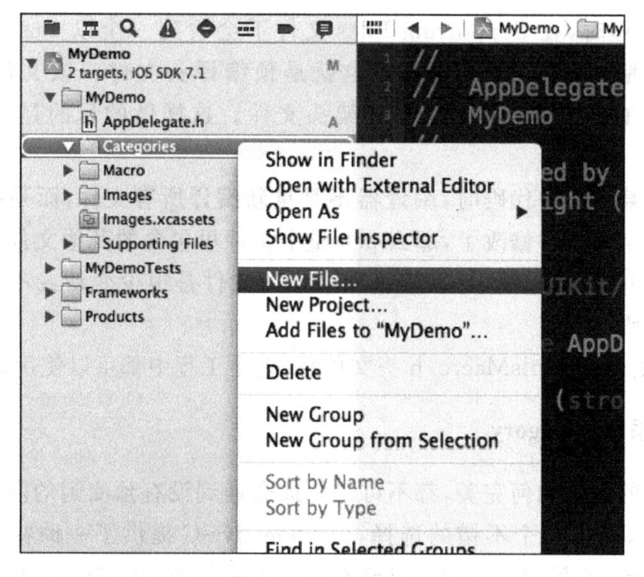

图 4.17 新建 file

(3) 选择 Cocoa Touch 下的 Objective-C category，如图 4.18 所示。

(4) 在 Category 中输入 Addition，在 Category on 中输入 UIView，如图 4.19 所示。

Catrgory on：指定为哪个类添加分类。

(5) 选择存储路径后，单击 Next 按钮，如图 4.20 所示。

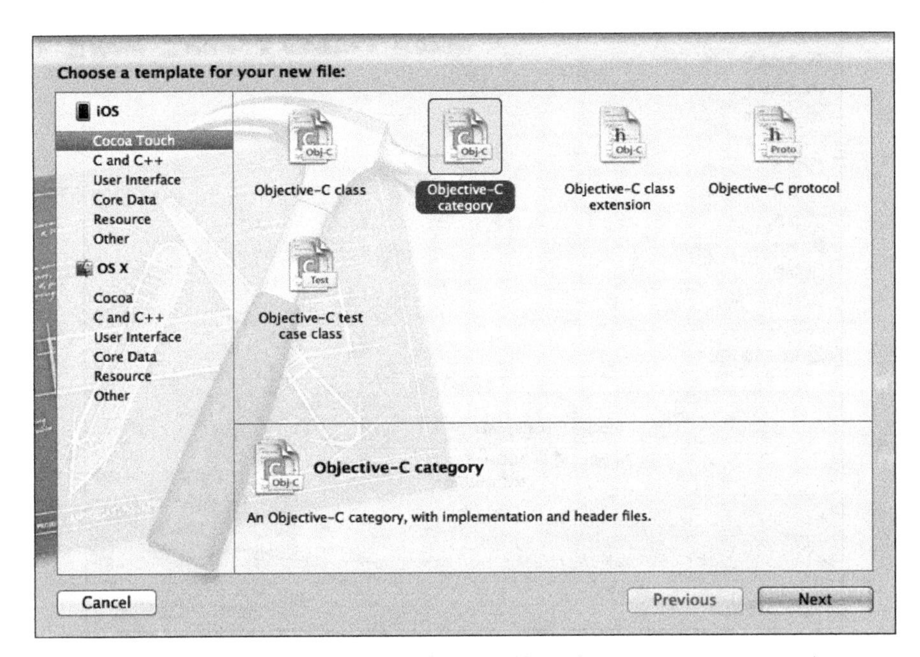

图 4.18 选择分类

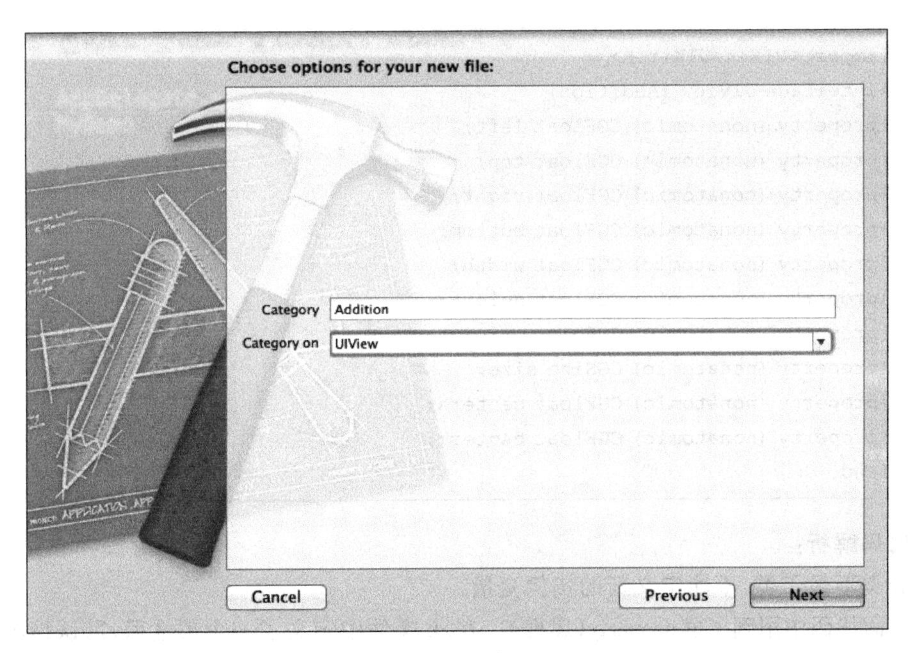

图 4.19 输入

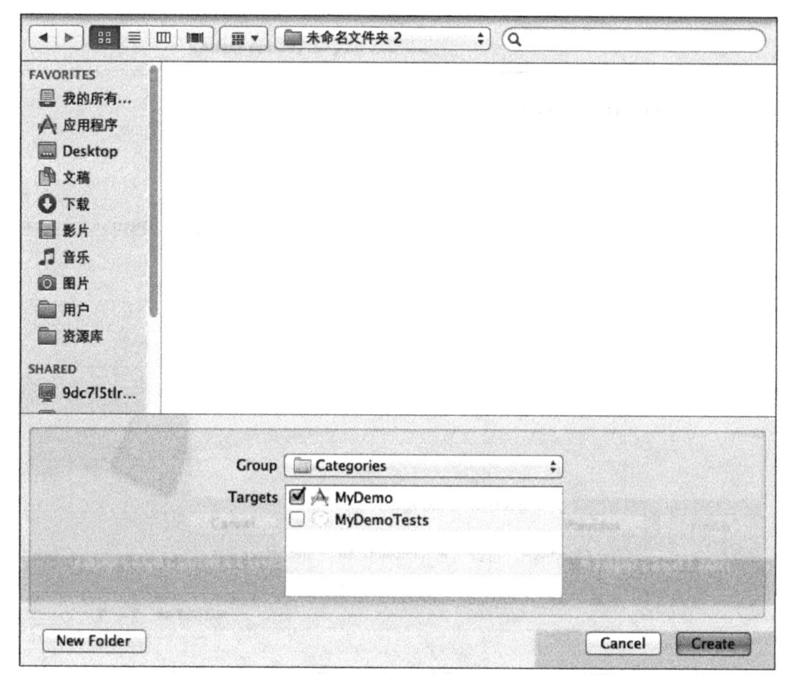

图 4.20　选择保存路径

(6) 单击 Categories 文件夹下的 UIView+Addition.h，修改代码如下：

```
#import<UIKit/UIKit.h>
@interface UIView (Addition)
@property (nonatomic) CGFloat left;
@property (nonatomic) CGFloat top;
@property (nonatomic) CGFloat right;
@property (nonatomic) CGFloat bottom;
@property (nonatomic) CGFloat width;
@property (nonatomic) CGFloat height;
@property (nonatomic) CGPoint origin;
@property (nonatomic) CGSize size;
@property (nonatomic) CGFloat centerX;
@property (nonatomic) CGFloat centerY;
@end
```

代码解析：

分类不像子类，不能定义新的实例变量。

上面代码中用到了@property(属性)。在.h 文件中定义了一个属性后，默认情况下系统会在.m 文件中实现该属性的 get 和 set 方法，但也可以根据需要重写 get 和 set 方法。

使用属性时，还需定义属性关键字。属性关键字的定义和使用如下：

atomic：原子操作(原子性是指事务的一个完整操作，操作成功就提交，反之就回滚。原子操作就是指具有原子性的操作)。在 Objective-C 属性设置里，默认的是 atomic。意思是 set/get 函数是一个原子操作，多线程同时调用 set 时，不会出现某一个线程执行完 set 的所有语句之前，另一个线程就开始执行 set 的情况。相当于在函数的头尾加了锁。这样做并发访问的效率会比较低。

nonatomic：非原子操作。一般不需要多线程支持时就用它，这样在并发访问的时候效率会比较高。在 Objective-C 里通常对象类型都应该声明为非原子性的。iOS 中程序启动的时候系统只会自动生成一个单一的主线程。程序在执行的时候，一般情况下是在同一个线程里对一个属性进行操作。如果在程序中我们确定某一个属性会在多线程中被使用，并且需要做数据同步，那么就必须设置成原子性的。若设置成非原子性的，就要在程序中用加锁之类的方式来实现数据同步。

readwrite：如果没有声明成 readonly，那就默认是 readwrite。可以用来赋值，也可以被赋值。

readonly：不可以被赋值。

assign：所有属性都默认为 assign，通常用于简单变量(如 int、float、CGRect 等)。一种典型情况是用在对象没有所有权的时候(如 delegate)，避免造成死循环(如果用 retain 会死循环)。

retain：释放旧对象。将旧对象的值赋予输入对象，再将输入对象的索引计数+1。属性必须是 Objective-C 对象，拥有对象所有权。

copy：建立一个相同对象。属性必须是 Objective-C 对象，一般常用于 NSString 类型。

iOS 5 推出了一个新的功能：ARC(Automatic Reference Counting)。简单地说，就是代码中自动加入了 retain/release，将原先需要手动添加的用来处理内存管理的引用计数的代码变为自动添加，一切由编译器完成。

下面是 ARC 中使用的关键字：

strong：相当于 retain。

weak：相当于 assign，不过对象在销毁之后会自动置为 nil，以防止野指针。

unsafe_unretained：不保留对象，只是简单引用。

(7) 单击 UIView+Addition.m，重写属性的 get 和 set 方法，代码如下：

```
#import "UIView+Addition.h"

@implementation UIView (Addition)
- (CGFloat)left {
    return self.frame.origin.x;
}
- (void)setLeft:(CGFloat)x {
    CGRect frame=self.frame;
    frame.origin.x=x;
    self.frame=frame;
}
```

```
- (CGFloat)top {
    return self.frame.origin.y;
}
- (void)setTop:(CGFloat)y {
    CGRect frame=self.frame;
    frame.origin.y=y;
    self.frame=frame;
}
- (CGFloat)right {
    return self.left+self.width;
}
- (void)setRight:(CGFloat)right {
    if(right==self.right){
        return;
    }
    CGRect frame=self.frame;
    frame.origin.x=right-frame.size.width;
    self.frame=frame;
}
- (CGFloat)bottom {
    return self.top+self.height;
}
- (void)setBottom:(CGFloat)bottom {
    if(bottom==self.bottom){
        return;
    }
    CGRect frame=self.frame;
    frame.origin.y=bottom-frame.size.height;
    self.frame=frame;
}
- (CGFloat)centerX {
    return self.center.x;
}
- (void)setCenterX:(CGFloat)centerX {
    self.center=CGPointMake(centerX, self.center.y);
}
- (CGFloat)centerY {
    return self.center.y;
}
- (void)setCenterY:(CGFloat)centerY {
    self.center=CGPointMake(self.center.x, centerY);
}
- (CGFloat)width {
```

```
    return self.frame.size.width;
}
- (void)setWidth:(CGFloat)width {
    CGRect frame=self.frame;
    frame.size.width=width;
    self.frame=frame;
}
- (CGFloat)height {
    return self.frame.size.height;
}
- (void)setHeight:(CGFloat)height {
    if(height==self.height){
        return;
    }

    CGRect frame=self.frame;
    frame.size.height=height;
    self.frame=frame;
}
- (CGPoint)origin {
    return self.frame.origin;
}
- (void)setOrigin:(CGPoint)origin {
    CGRect frame=self.frame;
    frame.origin=origin;
    self.frame=frame;
}
- (CGSize)size {
    return self.frame.size;
}
- (void)setSize:(CGSize)size {
    CGRect frame=self.frame;
    frame.size=size;
    self.frame=frame;
}
@end
```

代码解析：

重写属性的 get 和 set 方法，获取需要的数据。如果不重写，Xcode 会自动实现 get 和 set 方法。

(8) 单击 Supporting Files 文件夹下的 MyDemo-Prefix. pch，导入 UIView + Addition. h，如图 4.21 所示。

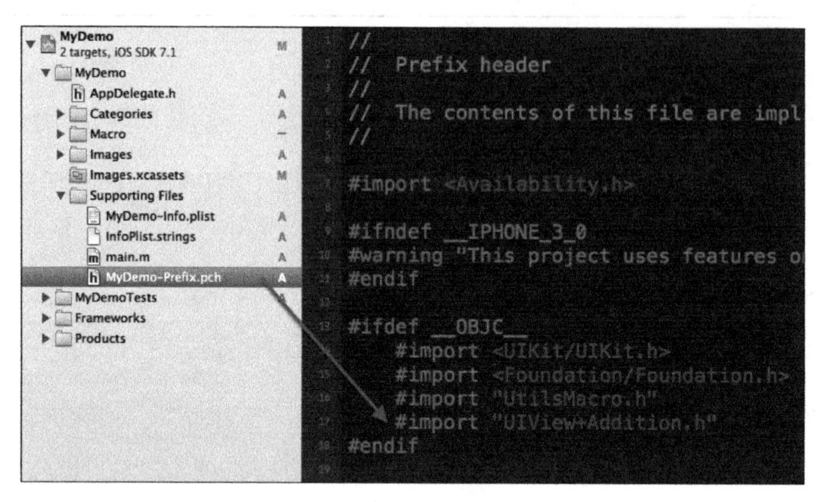

图 4.21 导入文件

5. 导入侧滑框架

自从 Facebook 使用了左右滑动菜单导航以后，国内外许多 App 都竞相模仿该功能。此功能有助于用户在不切换主界面的同时选择更多操作。这种左右滑动菜单可以将很多功能聚集在一起，让用户通过手势很方便地切换想要使用的功能，增强了用户体验。MFSideMenu 是 iOS 平台上一款非常不错的开源侧滑框架。

下载地址是：

```
https://github.com/mikefrederick/MFSideMenu
```

项目中会用到侧滑导航功能，因此先将侧滑框架导入到项目中。

(1) 将本书电子资源中的 MFSideMenu 文件夹拖放到工程 Venders 文件夹下，如图 4.22 所示。

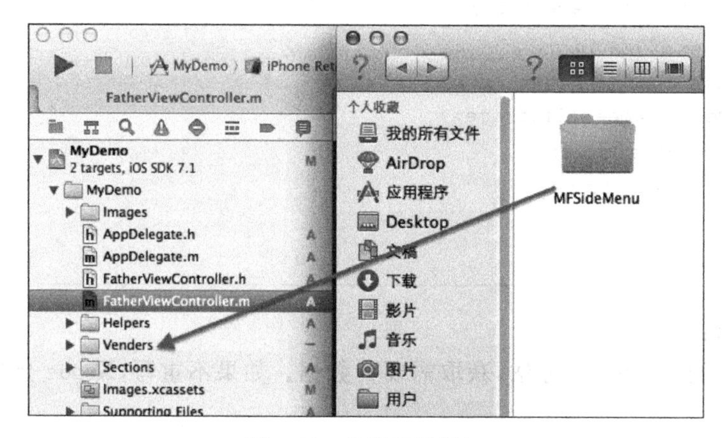

图 4.22 拖放侧滑框架

(2) 在弹出的对话框中选中 Create groups for any added folders，勾选 Copy items

into destination group's folder(if needed),在 Add to targets 中选择 MyDemo 之后单击 Finish 按钮,如图 4.23 所示。

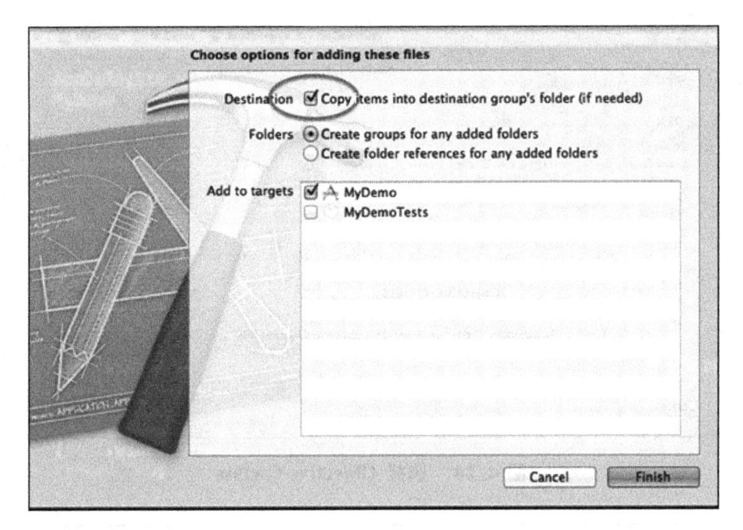

图 4.23 注意勾选

6. 为项目添加 Father 类

在项目开发中,通常会添加一个 Father 类并让其他子类继承这个父类,子类不仅能继承父类的方法和属性,还能重新编写父类方法。这就是面向对象的三大特性之一:继承。

继承是面向对象的一个核心概念。在 Objective-C 的继承体系中,位于最顶层的根类是 NSObject,定义的所有类都是它的子类。子类也叫扩展类或派生类。

继承使得子类可以从父类中获得一些属性和已有方法。要注意的是:如果子类中要直接使用父类继承过来的实例变量,那么必须将变量声明在接口部分中,而在实现部分声明的变量,子类无法继承使用。在实现部分声明和 synthesize 的实例变量都是私有的,子类不能直接访问,需要提供设置值和取值方法才可以访问这些变量。

下面创建一个 Father 类,作为 ViewController 的父类。

(1) 单击 AppDelegate.m,按下 command+N,选择 iOS 下的 Objective-C class,单击 Next 按钮,如图 4.24 所示。

(2) 在 Class 中填写 FatherViewController,在 SubClass of 中填写 UIViewController,此处不要勾选 Also create XIB file,单击 Next 按钮,如图 4.25 所示。

在 Xcode 中,创建视图有两种方式:使用 XIB 创建和使用纯代码创建。

使用 XIB 创建视图非常简单,只需简单用鼠标拖放就能完成视图的布局,不过对于一些复杂的视图,使用 XIB 创建会比较困难。

使用纯代码创建视图更加灵活,缺点是需要写相当多的代码,导致开发效率降低。由于 Father 类只需要让子类继承,因此不选择创建 XIB file。

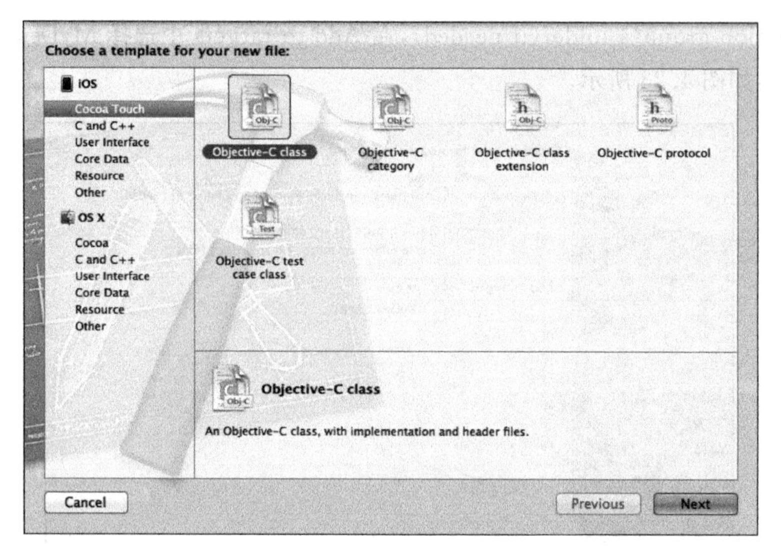

图 4.24　选择 Objective-C class

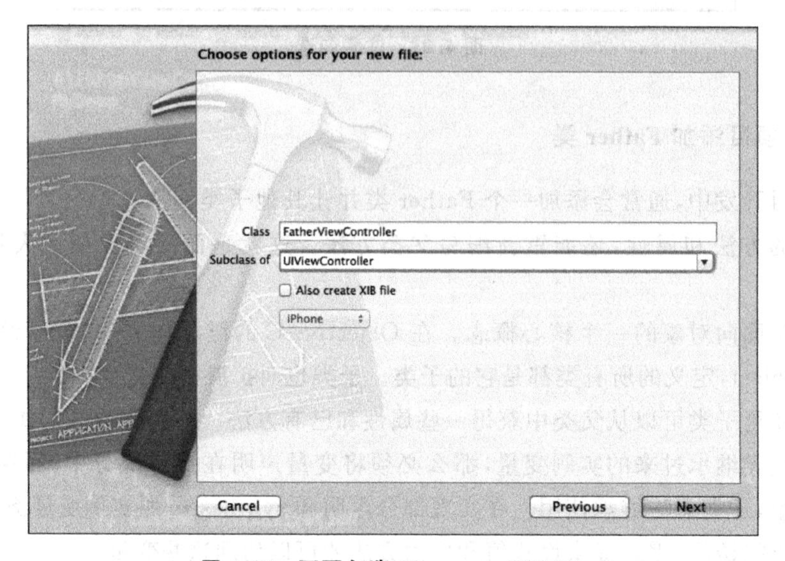

图 4.25　不要勾选 Also create XIB file

(3) 单击 FatherViewController.h,修改代码如下:

```
#import<UIKit/UIKit.h>
@interface FatherViewController: UIViewController
-(void)createLeftMenuBtn;                                    //创建左上角默认按钮
-(void)createLeftBtnWithFrame:(CGRect )rect Image:(UIImage * )image;
                                                             //创建左上角自定义按钮
-(void)createRightBtnWithFrame:(CGRect)rect Image:(UIImage * )image;
                                                             //创建右上角自定义按钮
-(void)leftBarChick;                                         //左上角自定义按钮单击事件
```

```
-(void)rightBarChick;                                //右上角自定义按钮单击事件
-(UIColor *)colorWithStr:(NSString *)hexStr;        //十六进制字符串转换颜色
@end
```

代码解析：

createLeftMenuBtn:方法——创建左上角默认菜单按钮，自动实现单击事件。

createLeftBtnWithFrame:Image:方法——创建左上角自定义按钮，需要重写 leftBarChick 方法。

createRightBtnWithFrame:Image:方法——创建右上角自定义按钮，需要重写 rightBarChick 方法。

colorWithStr:方法——十六进制字符串转换颜色。

(4) 单击 FatherViewController.m，修改代码如下：

```
#import "FatherViewController.h"
#import "MFSideMenuContainerViewController.h"  //导入侧滑框架头文件
@interface FatherViewController ()

@end
@implementation FatherViewController

-(void)viewDidLoad
{
    [super viewDidLoad];
    self.view.backgroundColor=[self colorWithStr:@"edf1f2"];
    if(IOS_7){
        self.edgesForExtendedLayout=0;
        //设置导航栏背景 高度为 64
        [self.navigationController. navigationBar setBackgroundImage: [UIImage
        imageNamed:@"navigationbar"] forBarMetrics:UIBarMetricsDefault];
    }else{
        //设置导航栏背景 高度为 44
        [self.navigationController. navigationBar setBackgroundImage: [UIImage
        imageNamed:@"navigationbar_44"] forBarMetrics:UIBarMetricsDefault];
    }
}
-(void)createLeftMenuBtn{
   UIButton *button=[UIButton buttonWithType:UIButtonTypeCustom];
   [button setFrame:CGRectMake(0.0f, 0.0f, 20.0f, 30.0f)];
   [button addTarget:self action:@selector(leftMenuChick) forControlEvents:
   UIControlEventTouchUpInside];
```

```
    [button setImage: [UIImage imageNamed: @" leftMenuBar "] forState:
    UIControlStateNormal];
    self.navigationItem. leftBarButtonItem = [[UIBarButtonItem alloc]
    initWithCustomView:button];
}
-(void)createLeftBtnWithFrame:(CGRect)rect Image:(UIImage *)image{
    UIButton *button=[UIButton buttonWithType:UIButtonTypeCustom];
    [button setFrame:rect];
    [button addTarget: self action: @selector (leftBarChick) forControl
    Events: UIControlEventTouchUpInside];
    [button setImage:image forState:UIControlStateNormal];
    self.navigationItem. leftBarButtonItem = [[UIBarButtonItem alloc]
    initWithCustomView:button];
}
-(void)createRightBtnWithFrame:(CGRect)rect Image:(UIImage *)image{
    UIButton *button=[UIButton buttonWithType:UIButtonTypeCustom];
    [button setFrame:rect];
    [button addTarget:self action:@selector(rightBarChick) forControlEvents:
    UIControlEventTouchUpInside];
    [button setImage:image forState:UIControlStateNormal];
    self.navigationItem. rightBarButtonItem = [[UIBarButtonItem alloc]
    initWithCustomView:button];
}

-(void)leftMenuChick{
    MFSideMenuContainerViewController *menuVC=(MFSideMenuContainerViewController
     *)self.navigationController.parentView Controller;
    [menuVC toggleLeftSideMenuCompletion:nil];
}
-(void)leftBarChick{

}
-(void)rightBarChick{

}
-(void)didReceiveMemoryWarning
{
    [super didReceiveMemoryWarning];
    // Dispose of any resources that can be recreated.
}

-(UIColor *)colorWithStr:(NSString *)hexStr
```

```
{
    long colorLong=strtoul([hexStr cStringUsingEncoding:NSUTF8StringEncoding],
    0,16);
    //通过位与方法获取三色值
    int R=(colorLong & 0xFF0000 )>>16;
    int G=(colorLong & 0x00FF00 )>>8;
    int B=colorLong & 0x0000FF;
    UIColor * GetColor=[UIColor colorWithRed:R/255.0 green:G/255.0 blue:B/
    255.0 alpha:1.0];
    return GetColor;

}
```

代码解析：

在iOS7中，苹果引入了一个新的属性，叫做setEdgesForExtendedLayout，它的默认值为UIRectEdgeAll。当容器是NavigationController时，默认的布局将从NavigationBar的顶部开始，所有UI元素上移44pt。可以设置edgesForExtendedLayout=0，不让UI元素上移。

在设置导航栏背景时，先判断用户当前设备的系统版本。如果系统版本大于7.0，那么就要加载一张高度为64pt的图片作为导航栏的背景图片，这是由于iOS 7的状态栏默认为透明，不占屏幕高度。如果系统版本小于7.0，则加载一张高度为44pt的图片，因为状态栏会占20pt高度。

leftMenuChick是调用侧滑框架中的方法，用来打开或关闭左侧滑页，因此需要导入侧滑头文件。

colorWithStr：方法是将传入的十六进制值转换为UIColor。很多时候，系统提供的颜色不能满足我们的需求，因此需要自定义颜色。也可将colorWithStr方式写成宏定义或者分类。

7. 创建注册页面

（1）右击FatherViewController.m，选择New Group，如图4.26所示。

（2）将其命名为Sections，并在其上右击，选择New File，如图4.27所示。

（3）选中iOS下的Cocoa Touch，再选择Objective-C class，单击Next按钮，如图4.28所示。

（4）在Class中输入RegisterViewController，在Subclass of中输入FatherViewController，勾选With XIB for user interface，如图4.29所示。

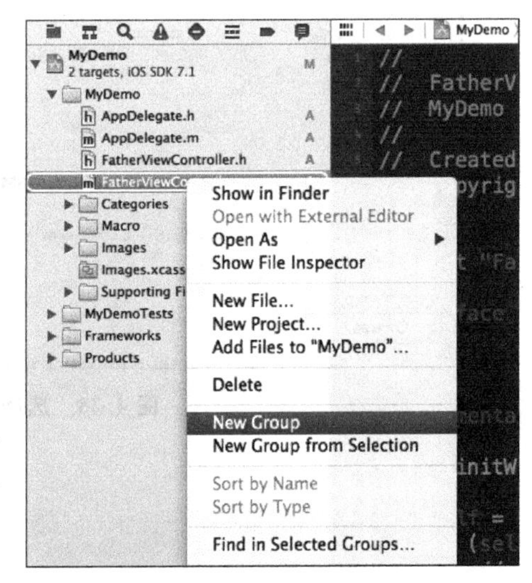

图4.26 新建Sections文件夹

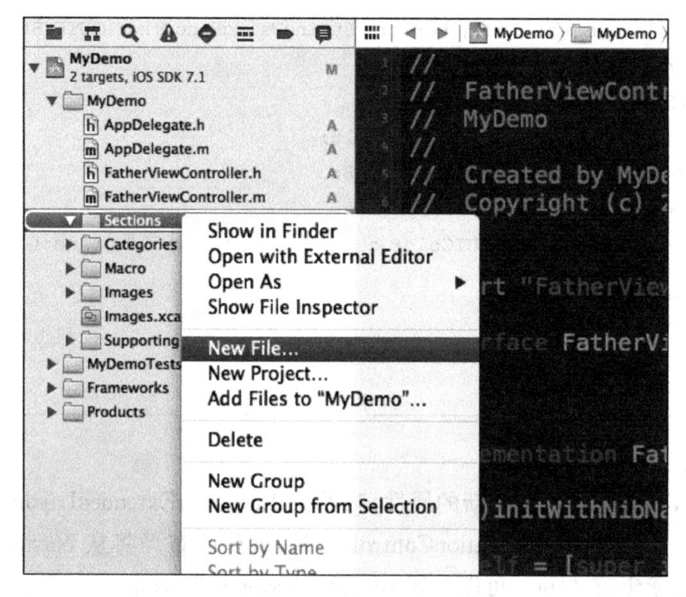

图 4.27　新建文件

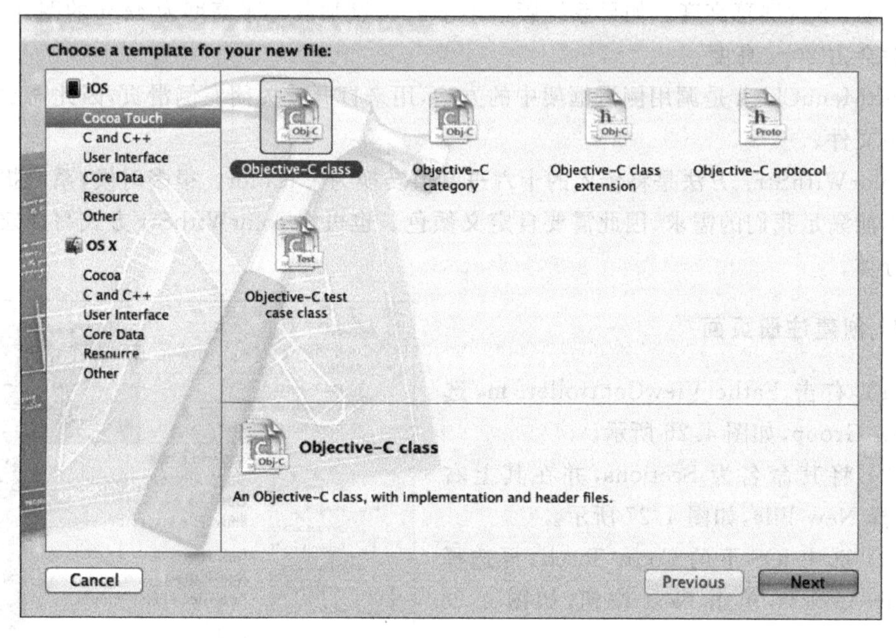

图 4.28　选择 Objective-C class

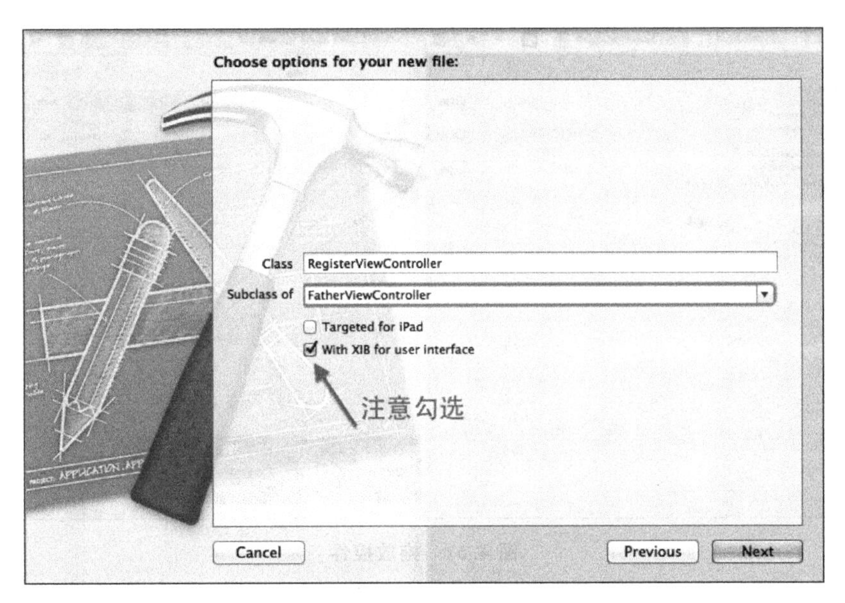

图 4.29 输入信息

这里让 RegisterViewController 继承 FatherViewController，这样 RegisterViewController 就能使用或重写父类的方法。由于 RegisterViewController 需要创建视图且视图比较简单，因此用 XIB 来创建视图。

(5) 选择项目保存地址后，完成添加，如图 4.30 所示。

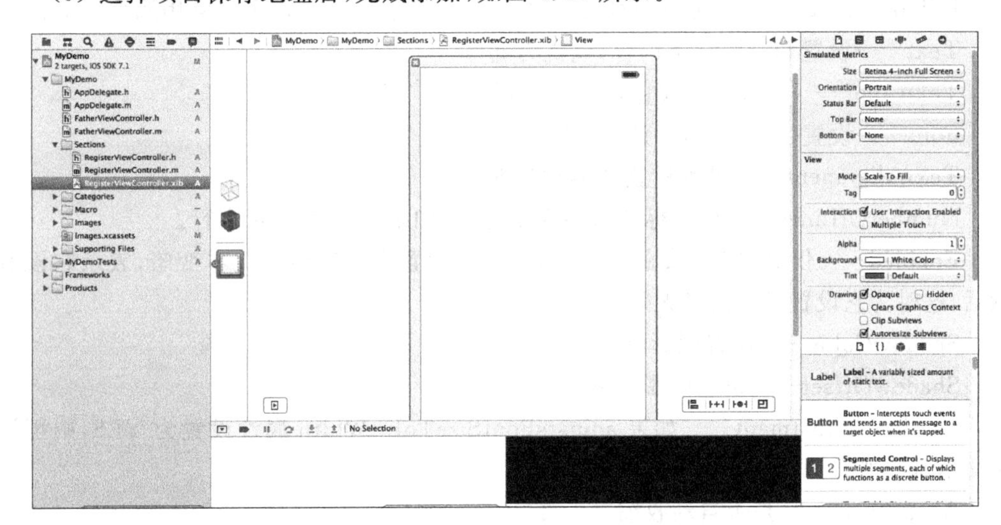

图 4.30 完成添加

8. 绘制注册页面界面

(1) 选中 RegisterViewController.xib，在右下角选中对象库并在下方输入框中输入 uilabel，拖动 3 个 UILabel 到 xib 界面上，如图 4.31 所示。

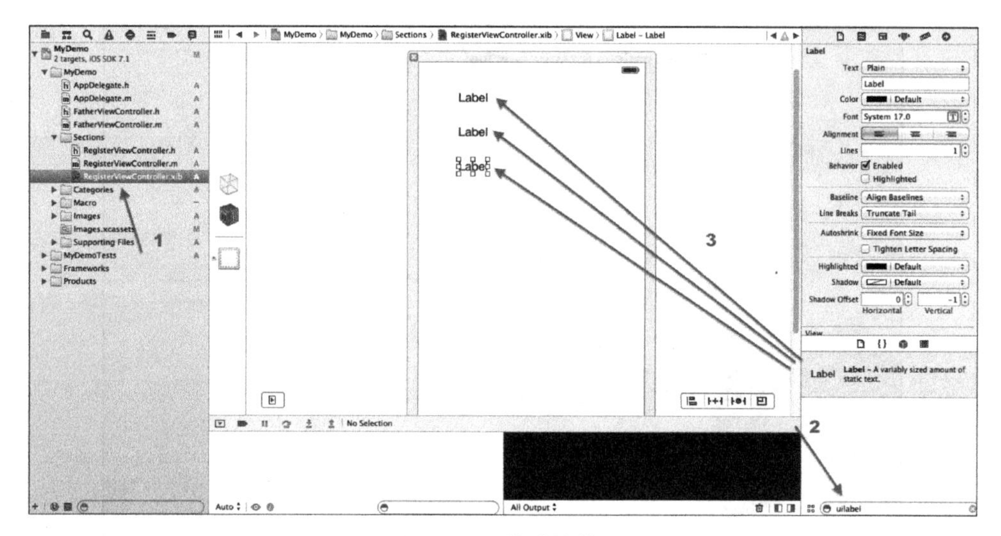

图 4.31　拖放控件

Xcode 工具区域右下角的 4 个选项依次为：

文件模板库(File Template Library)——包含文件模板。

代码片段库(Code Snippet Library)——包含代码片段。

对象库(Object Library)——包含各种可重用对象。

媒体库(Media Library)——包括用户所有的媒体文件。

UILable 是 iPhone 界面最基本的控件，主要用来显示文本信息。常用属性如下：

Text——设置和读取文本内容，默认为 nil。

TextColor——设置文字颜色，默认为黑色。

Font——设置字体大小，默认为 17。

TextAlignment——设置标签文本对齐方式。

NumberOfLines——设置标签最多显示行数；如果为 0，则表示多行。

Enabled——仅决定 Label 的绘制方式，将它设置为 NO 会使文本变暗，表示它没有激活，这时为它设置颜色值是无效的。

ShadowColor——设置阴影颜色。

ShadowOffset——设置阴影偏移量。

BaselineAdjustment——如果 adjustsFontSizeToFitWidth 属性设置为 YES，则可控制文本基线的行为。

Autoshrink——是否自动收缩。

MinimumScaleFactor——设置最小收缩比例，Label 宽度小于文字长度时，文字进行收缩，收缩超过比例后，停止收缩。

AdjustsLetterSpacingToFitWidth——改变字母的间距来适应 Label 的大小。

LineBreakMode——设置文字过长时的显示格式。

AdjustsFontSizeToFitWidth——设置字体大小适应 label 的宽度。

(2) 单击界面上的第 1 个 UILabel，设置 Text 属性为“账号：”，Color 属性为 Dark

Gray Color,Font 属性为 System 13.0,如图 4.32 所示。

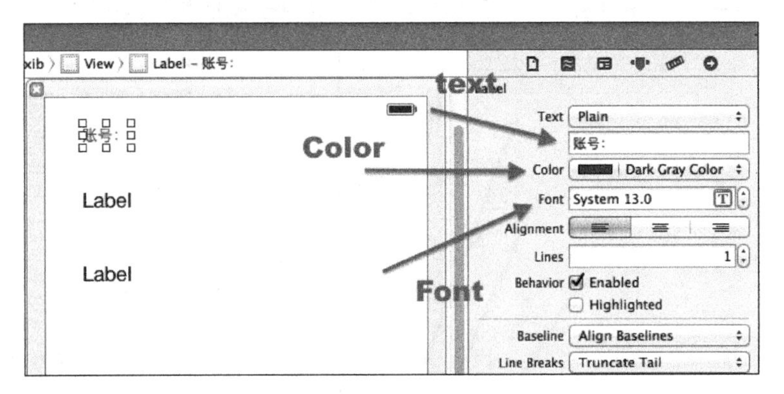

图 4.32 修改属性

图 4.32 所示区域为属性编辑区,会根据不同控件显示不同属性,能非常方便地设置控件的常用属性,如 UILabel 的颜色、对齐方式、字体属性等。

(3) 设置剩下的两个 UILabel。Text 属性分别为“密码:”和“重复密码:”,Color 属性均为 Dark Gray Color,Font 属性都为 System 13.0,如图 4.33 所示。

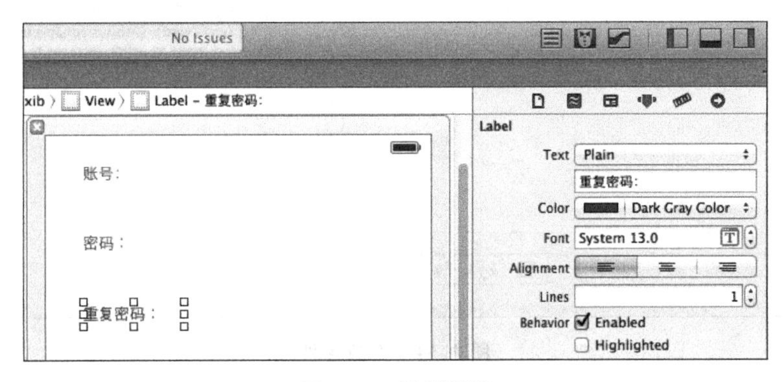

图 4.33 设置属性

(4) 在右下角输入 uitextfield,拖动 3 个 UITextField 到 xib 界面上并调整位置,如图 4.34 所示。

UITextField 主要用来接收用户的输入信息。常用的属性如下:

TextColor——设置文字颜色。

BackgroundColor——设置文本框背景颜色。

BorderStyle——设置边框的风格。

TextAlignment——设置文字对齐方式。

Font——设置文字大小和字体。

AdjustsFontSizeToFitWidth——设置自适应文本框大小。

ClearsOnBeginEditing——设置是否拥有一键清除的功能。

ClearButtonMode——设置一键清除按钮是否出现。

Placeholder——设置初始隐藏文字。

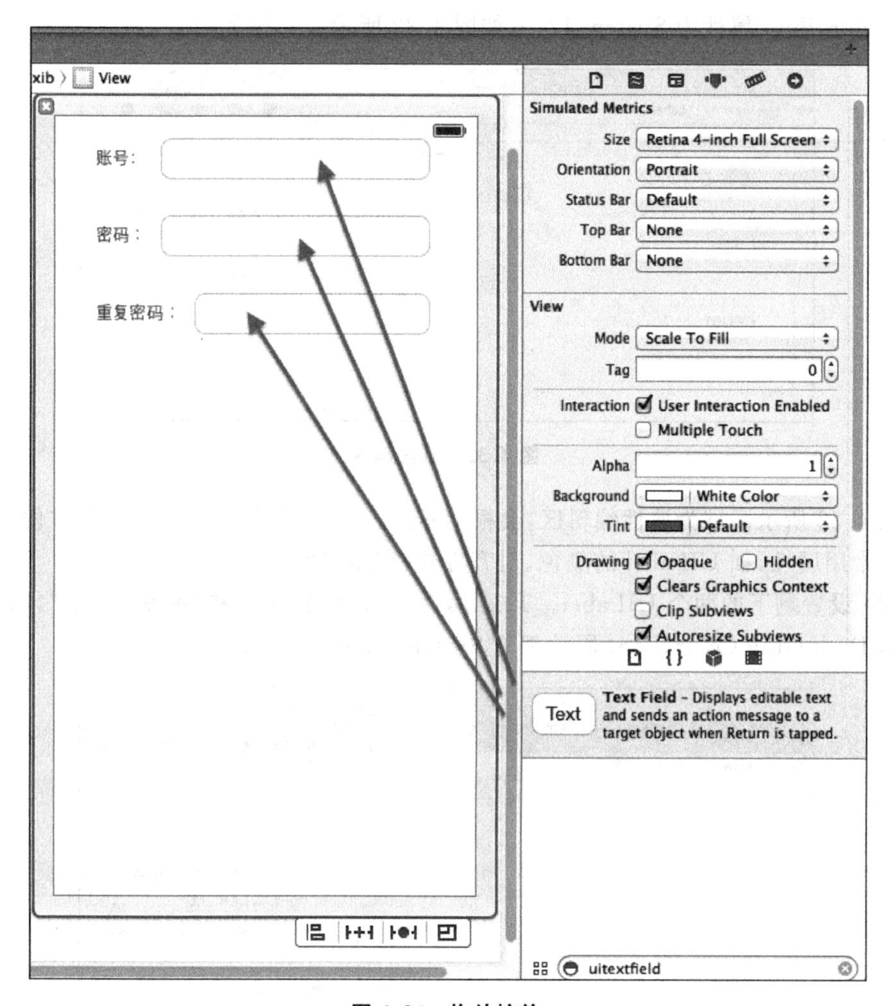

图 4.34 拖放控件

Background——当 UITextField 的样式为 UITextBorderStyleNone 时，修改背景图片。

LeftView——设置左边 view。

LeftViewMode——设置左边 view 的出现模式。

RightView——设置右边 view。

RightViewMode——设置右边 view 的出现模式。

ContentVerticalAlignment——设置文本垂直方向的对齐方式。

(5) 单击第 1 个 UITextField，设置 Color 属性为 Light Gray Color，Font 属性为 System 12.0，Placeholder 属性为“请输入您的账号”，在 Border Style 中选择第 1 个，如图 4.35 所示。

(6) 设置剩下的两个 UITextField，设置 Color 的属性都为 Dark Gray Color，Font 属性都为 System 12.0，Placeholder 的属性分别设置为“密码需要六位以上”“请保持密码一致”，在 Border Style 中都选择第 1 个，如图 4.36 所示。

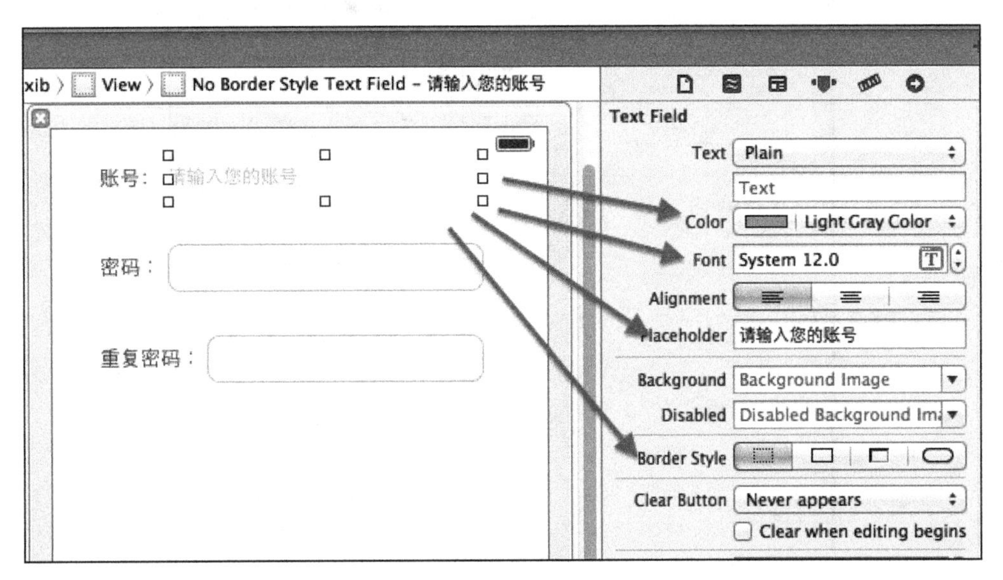

图 4.35　设置属性

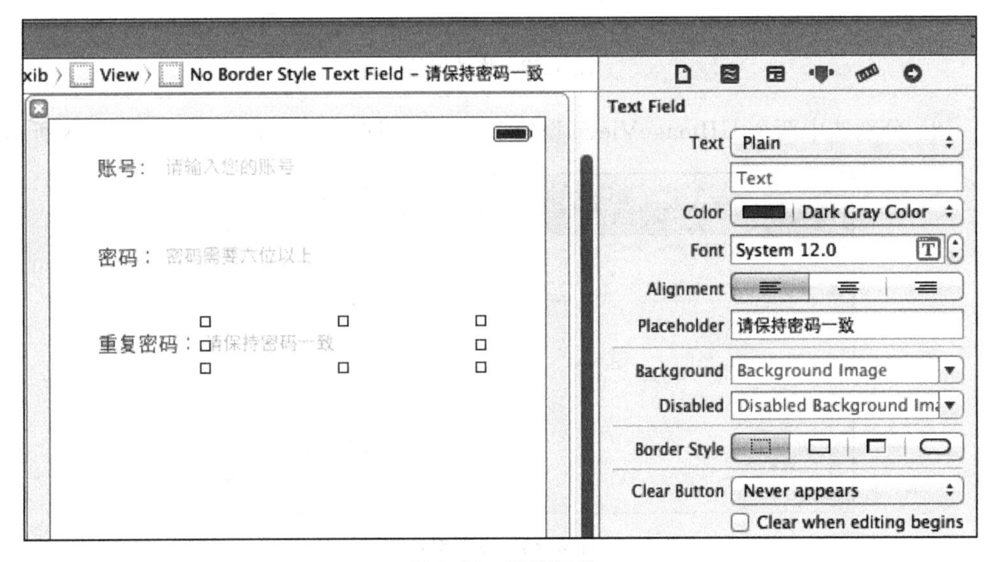

图 4.36　设置属性

(7) 在右下角输入 uiimageview，拖动两个 UIImageView 到 xib 界面上并调整位置，如图 4.37 所示。

UIImageView 的使用频率非常高，主要用来加载图片。常用的属性是 UIImage，为 UIImageVIew 指定一个 Image 图像。

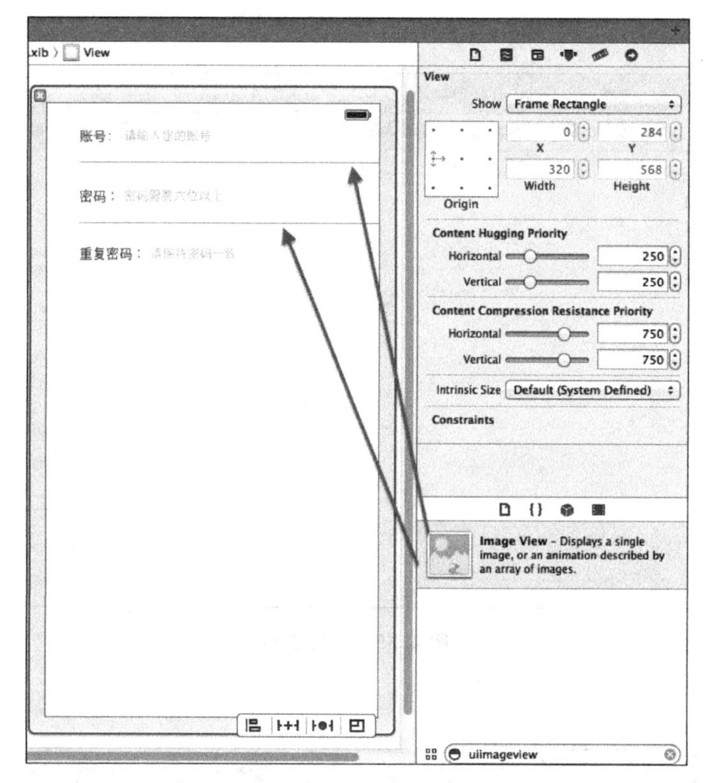

图 4.37 拖放控件

(8) 分别选中两个 UIImageView，设置 Image 的属性都为 line. png，如图 4.38 所示。

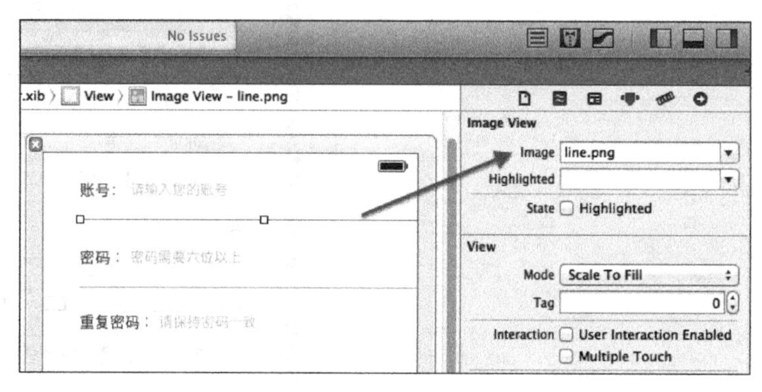

图 4.38 设置图片

(9) 在右下角输入 uibutton，拖动 1 个 UIButton 到 xib 界面上，设置 Title 属性为“注册”，TextColor 属性为 White Color，Background 属性为 checkopinion. png，如图 4.39 所示。

9. 关联注册界面上的控件

(1) 单击 Xcode 右上角的按钮，弹出助手面板，确认助手面板中为 RegisterView Controller. h，选中第 1 个 UITexiField。右击，拖动弹出面板中的 New Referencing Outlet 右边的加号到助手面板中，如图 4.40 所示。

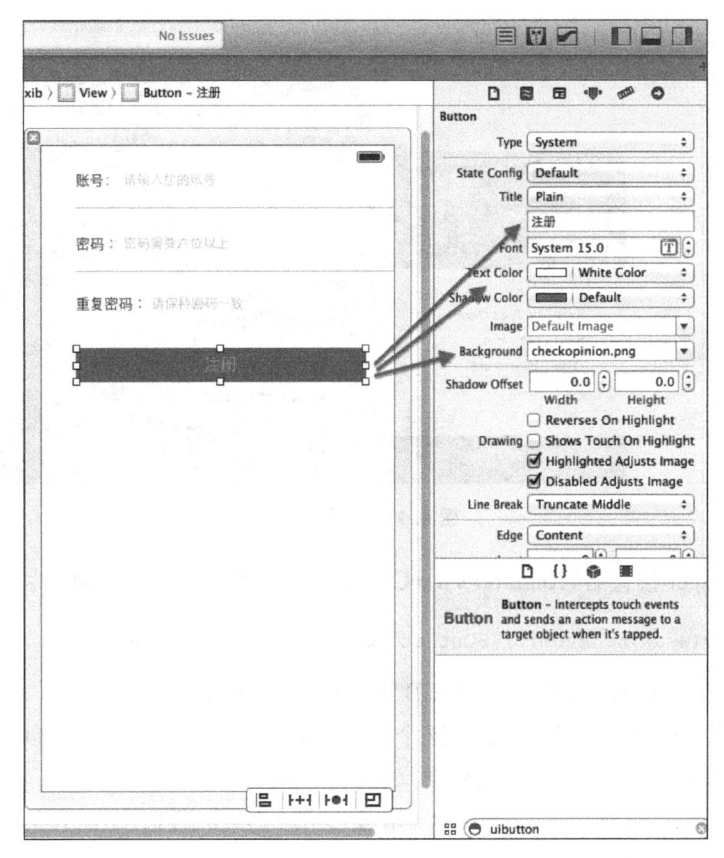

图 4.39　设置属性

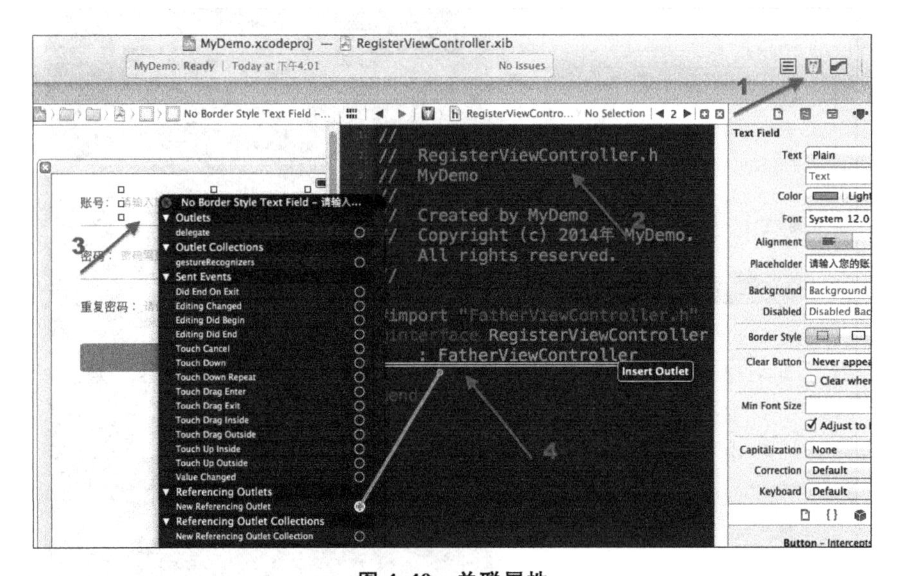

图 4.40　关联属性

(2) 在弹出框的 Name 项中输入 username，其他地方不用更改，单击 Connect，如

图 4.41 所示。

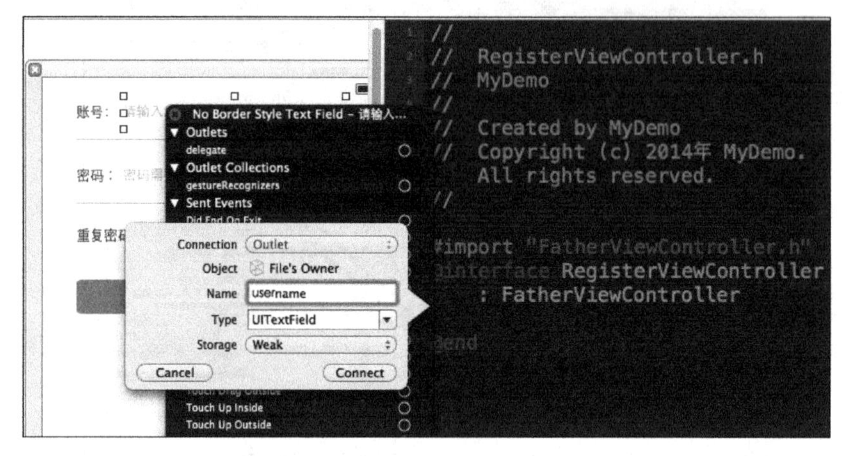

图 4.41 属性命名

单击 Connect 按钮后，RegisterViewController.h 就增加了一个属性：

```
@property(weak,nonatomic)IBOutlet UITextField * username;
```

这样就能通过 username 属性来控制 XIB 上对应的 UITextField 控件。

注意到 IBOutlet 关键字，它表示这个对象是在 Interface Builder 中创建的。对于编译器而言，这只是一个标记，没有其他含义。

(3) 关联剩下的两个 UITextField，重复步骤(2)，分别命名为 password、password Repeat，如图 4.42 所示。

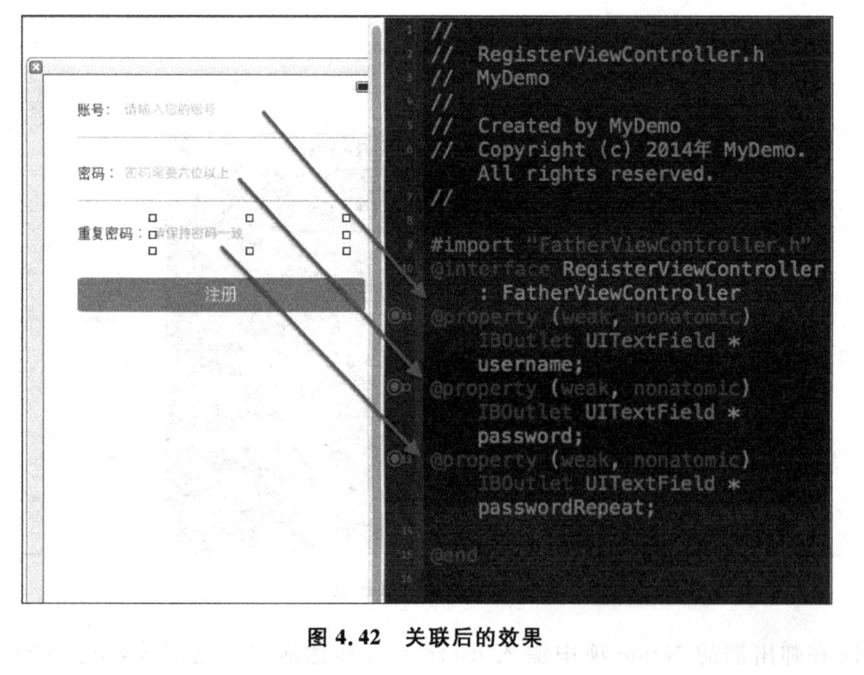

图 4.42 关联后的效果

（4）选中 xib 上的“注册”按钮后右击，拖动弹出层上 Touch Up Inside 右边的圆圈到助手面板上，如图 4.43 所示。

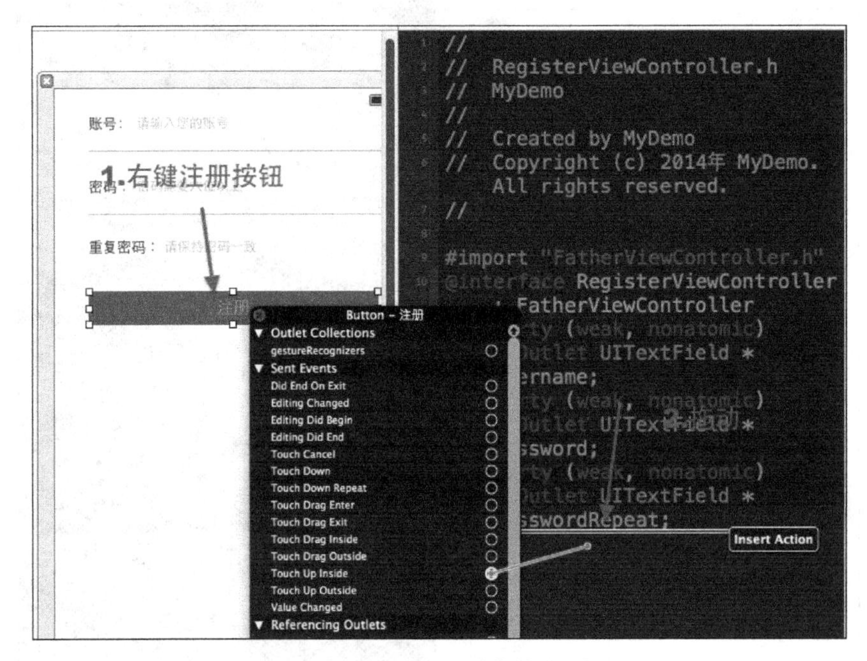

图 4.43 关联事件

这里为 Button 关联 Touch Up Inside 事件，也就是单击 Button 时触发的事件，另外，还可以关联其他事件，比如 Touch Down(按下)事件等。

（5）在弹出框的 Name 框中输入 registChick，其他地方不用更改，单击 Connect 按钮，如图 4.44 所示。

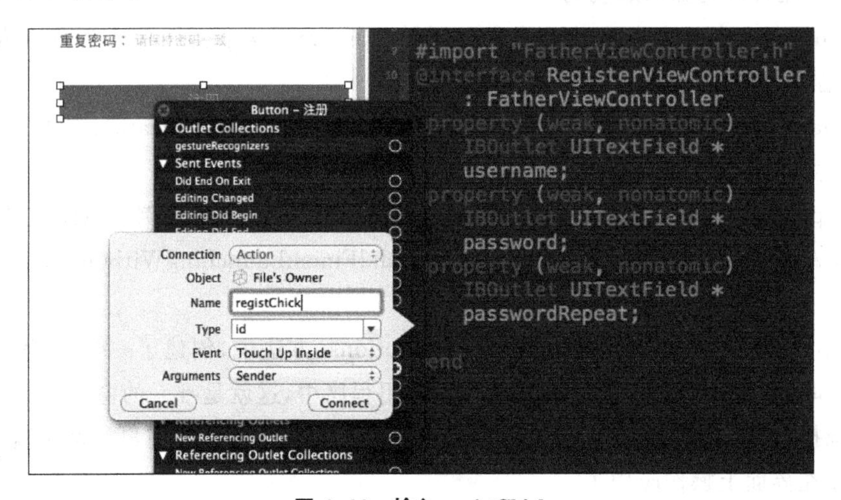

图 4.44 输入 registChick

（6）最终效果如图 4.45 所示。

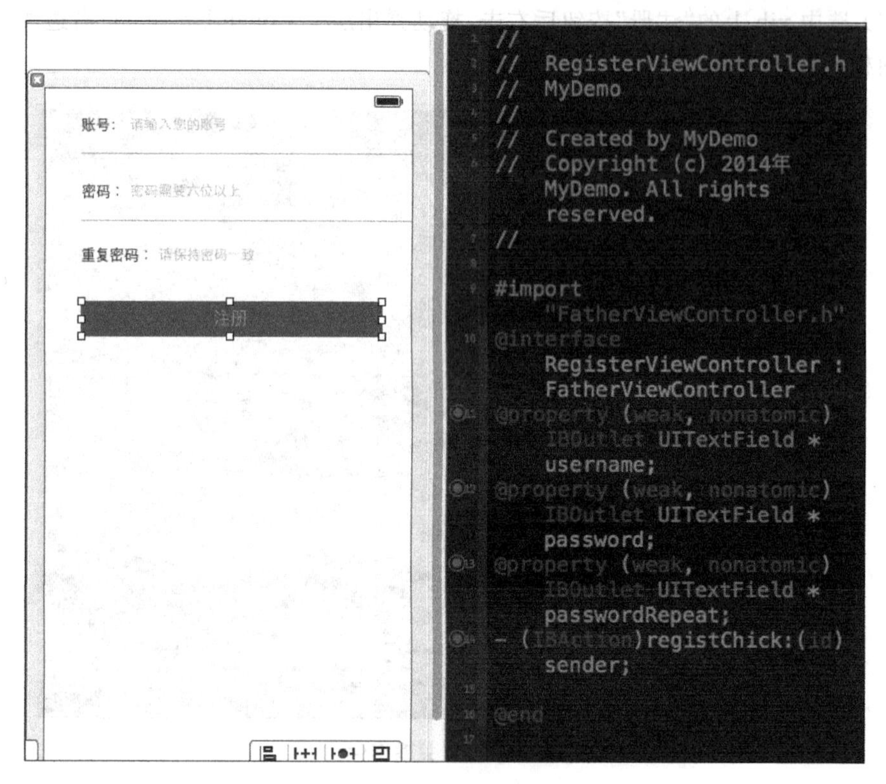

图 4.45　关联后的最终效果

10. 设置注册页面为启动页面

iOS 应用程序是如何启动的？

在 Supporting 文件夹下，有一个文件叫做 main. m。这个文件是整个应用执行的入口，随着程序的启动，main 方法被调用。

在 main 方法中调用 UIApplicationMain，从而创建了 UIApplication 类的一个实例，这是一个共享的应用实例。

UIApplicationMain 创建了应用程序委托(AppDelegate)的实例。系统完成了内部设置后，AppDelegatede 实例的 application：didFinishLaunchingWithOptions：方法被调用。

在 application：didFinishLaunchingWithOptions：方法中，创建了一个 UIWindow 类的实例，即显示给用户的界面。在一个 iOS 应用程序中，这就是唯一的窗口实例。

系统执行到[self. window makeKeyAndVisible]方法后，整个用户界面呈现出来，用户就可以在界面上操作应用了。

因此我们要设置程序的启动页面，需要到 application：didFinishLaunchingWithOptions：中添加代码。

(1) 单击 AppDelegate. m 文件，引入 RegisterController. h 头文件，如图 4.46 所示。

```
#import "AppDelegate.h"
#import "RegisterViewController.h"
@implementation AppDelegate
```

图 4.46 引入头文件

(2) 重写 application:didFinishLaunchingWithOptions:方法,添加代码如下:

```
-(BOOL)application:(UIApplication *)application didFinishLaunchingWithOptions:
(NSDictionary *)launchOptions
{
    self.window=[[UIWindow alloc] initWithFrame:[[UIScreen mainScreen]
    bounds]];
    //Override point for customization after application launch.
    RegisterViewController * registerVC=[[RegisterViewController alloc]
    initWithNibName:@"RegisterViewController" bundle:nil];
    self.window.rootViewController=registerVC;
    self.window.backgroundColor=[UIColor whiteColor];
    [self.window makeKeyAndVisible];
    return YES;
}
```

代码解析:

application:didFinishLaunchingWithOptions:方法——iOS 程序启动时会调用此方法,其中第二个参数 launchOptions 为 NSDictionary 类型的对象,里面存储了此程序启动的原因。例如,若用户直接启动,lauchOptions 内无数据。若由其他应用程序通过 openURL:启动,则 lauchOptions==UIApplicationLaunchOptionsURLKey 对应的对象为启动 URL(NSURL), lauchOptions==UIApplicationLaunchOptionsSourceApplicationKey 对应启动的源应用程序的 Bundle ID。

(3) 在顶部工具栏中,找到 Scheme 按钮,在该按钮的下拉菜单中选择 iPhone Retina (3.5-inch),如图 4.47 所示。

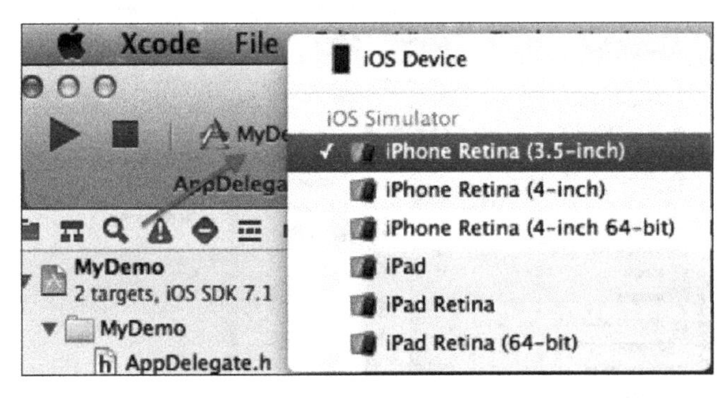

图 4.47 修改项目的 Scheme

(4) 单击工具栏左上角的▶按钮(快捷键 command+R),编译并运行程序,模拟器打

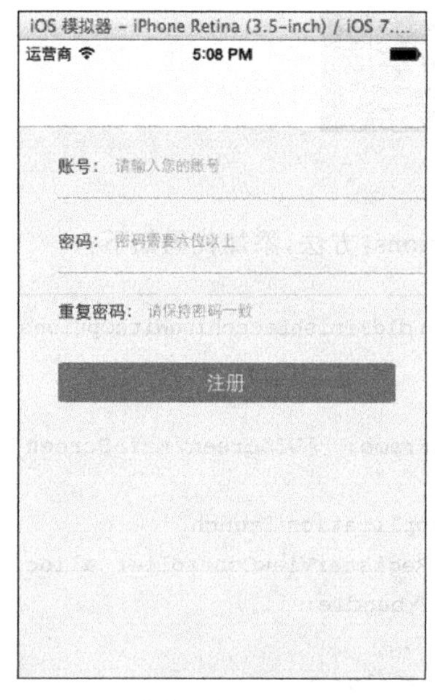

图 4.48 在模拟器中运行应用程序

开界面如图 4.48 所示。

4.4.2 客户端与服务端交互

客户端与服务端交互，使用 HTTP 协议是最简单易用的方法。完整的过程大致如下：

（1）客户端通过 HTTP 协议访问服务端的某个页面。

（2）服务端经过处理后，把这个页面生成为仅有数据的文本流。

（3）客户端得到文本流后，反序列化为对象并显示。

下面介绍一个 iOS 平台上非常棒的第三方开源框架来实现数据的请求与传递，它就是 AFNetWorking。

AFNetworking 适用于 iOS 以及 Mac OS X。它构建于 NSURLConnection、NSOperation，以及其他熟悉的 Foundation 技术之上。它拥有良好的架构、丰富的 API 以及模块化构建方式，使得开发者使用起来非常轻松。

AFNetworking 的下载地址：

```
https://github.com/AFNetworking/AFNetworking/
```

1. 添加 AFNetworking

（1）右击 FatherViewController.m，选择 New Group，将其命名为 Venders，将本书电子资源中的 AFNetWorking 文件夹拖曳进去，如图 4.49 所示。

图 4.49 导入 AFNetworking 框架

(2) 勾选 Copy items into destination group's folder(if needed),如图 4.50 所示。

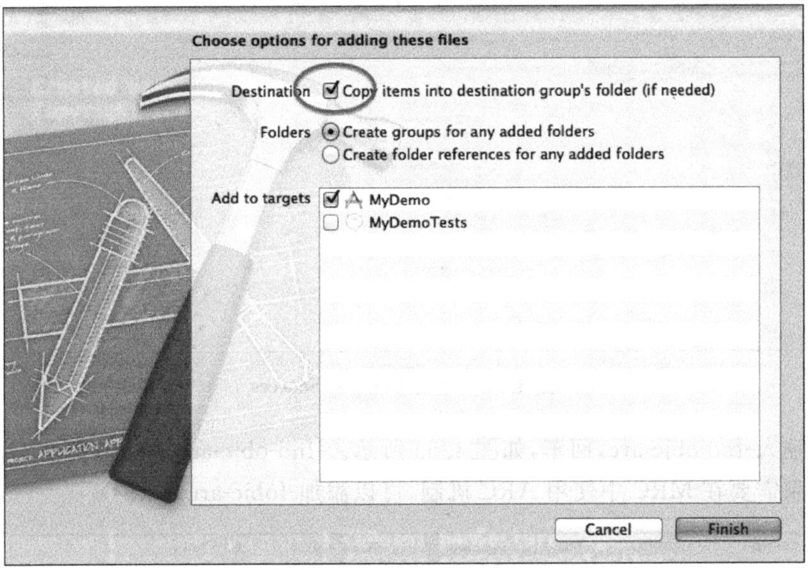

图 4.50 注意勾选

(3) 按下 command+B 编译工程,此时会报告许多错误,如图 4.51 所示。

这是由于引入的 AFNetWorking 是基于 MRC(手动内存管理),而工程用到了 ARC(自动内存管理),因此需要对 AFNetWorking 报错文件进行设置,从而使 Xcode 不用 ARC 模式来编译 AFNetWorking 文件。

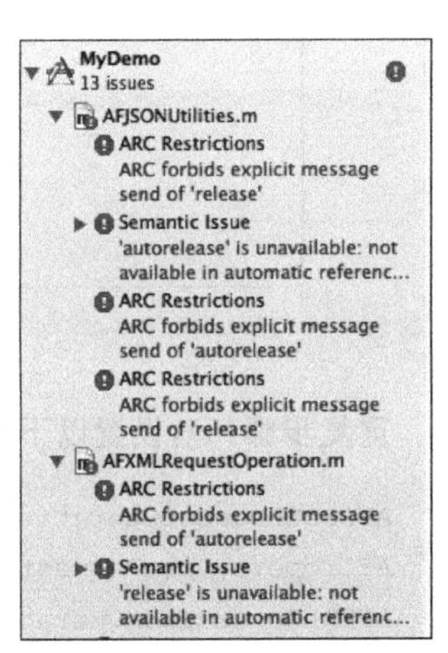

图 4.51 编译报错

在 MRC 的内存管理模式下,与变量管理相关的方法有 retain、release 和 autorelease。retain 和 release 方法操作的是引用记数,当引用记数为零时,便自动释放内存,并且可以用 NSAutoreleasePool 对象对加入自动释放池(autorelease 调用)的变量进行管理,当用尽时回收内存。

ARC 是新 LLVM 3.0 编译器的特性,完全摆脱了手动内存管理的烦琐工作。在项目中使用 ARC 非常简单,所有的编程都和以前一样,只是不再调用 retain、release、autorelease。启用 ARC 之后,编译器会自动在适当的地方插入 retain、release、autorelease 语句。开发者不再需要担心内存管理,因为编译器会处理一切事情。注意 ARC 是编译器的特性,而不是 iOS 运行时特性(除了 weak 指针系统),它也不是其他语言中的垃圾收集器。因此 ARC 和 MRC 的性能是一样的,有些时候还能更加快速,因为编译器还可以实现某些优化。

(4) 依次单击工程→MyDemo(在 TARGETS 下)→Build Phases→Compile Sources,如图 4.52 所示。

(5) 在上方搜索框中输入错误提示文件 AFJSONUtilities.m,双击搜索结果,然后在

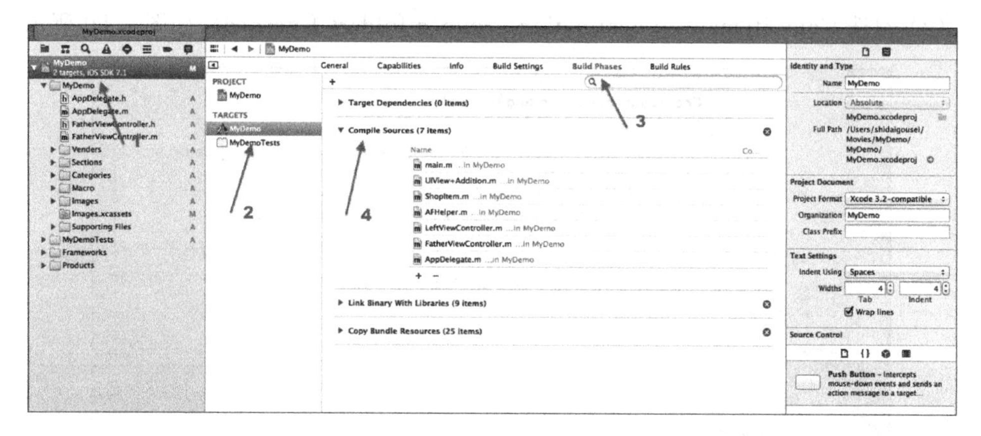

图 4.52　单击 Compile Sources

输入框中输入-fno-objc-arc，回车，如图 4.53 所示。-fno-objc-arc 表示该文件不使用 ARC 机制。如果需要在 MRC 中使用 ARC 机制，可以添加-fobjc-arc。

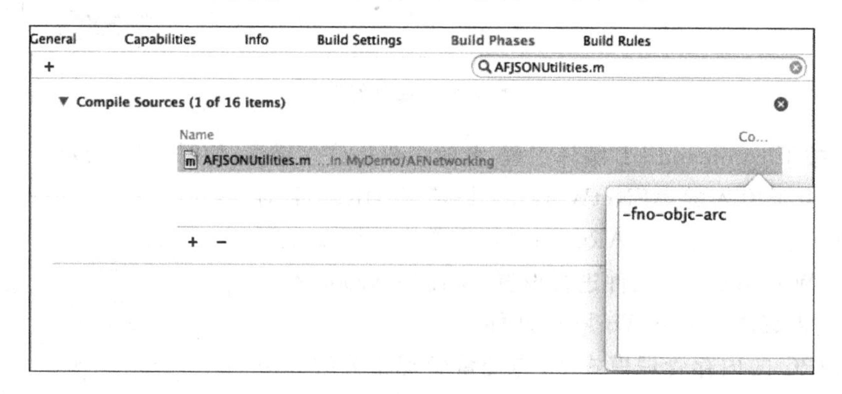

图 4.53　为文件设置-fno-objc-arc

重复步骤(5)，设置以下剩余报错文件：

```
AFJSONRequestOperation.m
AFPropertyListRequestOperation.m
AFURLConnectionOperation.m
UIImageView+AFNetworking.m
AFImageRequestOperation.m
AFNetworkActivityIndicatorManager.m
AFXMLRequestOperation.m
AFHTTPRequestOperation.m
AFHTTPClient.m
```

全部设置完成后，按下 command＋N，就会提示 Build Successed 了。

2. 编写下载助手类

通常，开发者需要为项目编写下载助手类，目的是方便与服务端进行数据传递，减少代码冗余。

(1) 右击 FatherViewController.m,选择 New Group,新建文件夹 Helpers。

(2) 选中 Helpers 文件夹,按下 command+N,添加一个 Objective-C class,如图 4.54 所示。

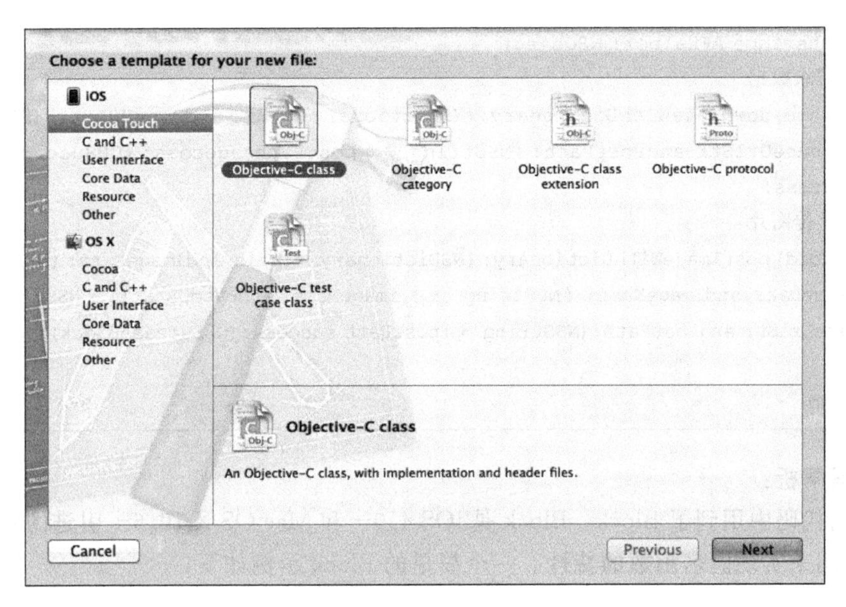

图 4.54 新建 Objective-C class

(3) 在 Class 中输入 AFHelper,在 Subclass of 中输入 NSObject,单击 Next,如图 4.55 所示。

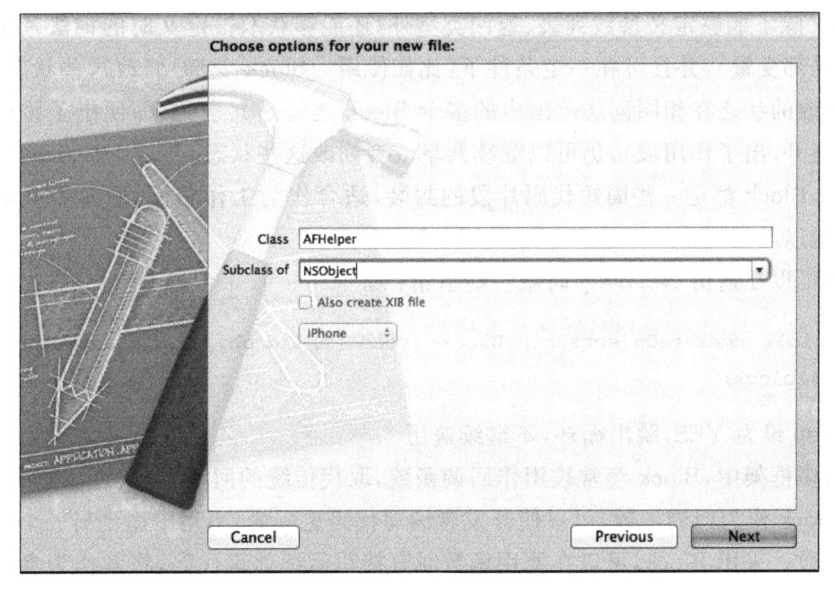

图 4.55 输入信息

(4) 单击 AFHelper.h,修改代码如下:

```
#import<Foundation/Foundation.h>
typedef void(^CBSuccessBlock)(NSDictionary *dic);

@interface AFHelper: NSObject
//请求数据
+(void)downDataWithDictionary:(NSDictionary *)dic andBaseURLStr:(NSString
*)baseUrlStr andPostPath:(NSString *)postPath success:(CBSuccessBlock)
success;
//上传图片
+(void)postImageWithDictionary:(NSDictionary *)dic andImageData:(NSData *)
imageData andImageName:(NSString *) imageName andBaseURLStr:(NSString *)
baseUrlStr andPostPath:(NSString *)postPath success:(CBSuccessBlock)success;

@end
```

代码解析:

上面代码中用到了 Block。Block 是 iOS 4.0+和 Mac OS X 10.6+引进的对 C 语言的扩展,用来实现匿名函数的特性。一个简单的 Block 示例如下:

```
int (^maxBlock)(int, int)=^(int x, int y) {
  return x>y ? x: y;
};
```

Block 除了能够定义参数列表、返回类型外,还能够获取被定义时的词法范围内的状态(比如局部变量),并且可在一定条件下(比如使用__block 变量)修改这些状态。此外,这些可修改的状态在相同词法范围内的多个 Block 之间是共享的,即便出了该词法范围(比如栈展开、出了作用域),仍可以继续共享或者修改这些状态。

通常,Block 都是一些简短代码片段的封装,适合作为工作单元,通常用来实现并发任务、遍历以及回调。

比如可以在遍历 NSArray 时做一些事情:

```
-(void)enumerateObjectsUsingBlock:(void (^)(id obj, NSUInteger idx, BOOL *
stop))block;
```

其中将 stop 设为 YES,跳出循环,不继续遍历。

在很多框架中,Block 经常被用作回调函数,取代传统的回调方式。

用 Block 作为回调函数,可以使开发者编写代码更顺畅,不用中途到另一个地方写一个回调函数。采用 Block,可以在调用函数时直接写后续处理代码,将其作为参数传递过去,供任务执行结束时回调。

(5) 单击 AFHelper.m,修改代码如下:

```
#import "AFHelper.h"
#import "AFNetworking.h"
@implementation AFHelper

+(void)downDataWithDictionary:(NSDictionary *)dic andBaseURLStr:(NSStr-
ing *)baseUrlStr andPostPath:(NSString *)postPath success:(CBSuccessBlock)
success {
    NSURL *baseUrl=[NSURL URLWithString:baseUrlStr];
    AFHTTPClient *aClient=[AFHTTPClient clientWithBaseURL:baseUrl];
    [aClient postPath: postPath parameters: dic success: ^ (AFHTTPRequ -
estOperation *operation, id responseObject) {
        NSDictionary *dic=[NSJSONSerialization JSONObjectWithData:resp-
onseObject options:1 error:nil];
        success(dic);
    } failure:^(AFHTTPRequestOperation *operation, NSError *error) {
        UIAlertView *alert=[[UIAlertView alloc]initWithTitle:@"提示" message:
@"网络连接失败" delegate:nil cancelButtonTitle:nil otherButtonTitles:@"好的",
nil];
        [alert show];
    }];
}

+(void)postImageWithDictionary:(NSDictionary *)dic andImageData:(NSData *)
imageData andImageName: (NSString *) imageName andBaseURLStr: (NSString *)
baseUrlStr andPostPath:(NSString *)postPath success:(CBSuccessBlock)success{
    AFHTTPClient * aClient = [AFHTTPClient clientWithBaseURL: [NSURL
URLWithString:baseUrlStr]];
    NSMutableURLRequest * fileUpRequest = [aClient multipartFormRequestWith
Method:@"POST" path:postPath parameters:dic constructingBodyWithBlock:^(id<
AFMultipartFormData>formData) {
        [formData appendPartWithFileData:imageData name:@"file" fileName:image
Name mimeType:@"image/png"];
    }];
    AFHTTPRequestOperation * op = [[AFHTTPRequestOperation alloc] initWith -
Request:fileUpRequest];
    [op start];
    [op setCompletionBlockWithSuccess:^(AFHTTPRequestOperation *operation, id
responseObject) {
        NSDictionary * dic = [NSJSONSerialization JSONObjectWithData: response
Object options:1 error:nil];
        success(dic);
    } failure:^(AFHTTPRequestOperation *operation, NSError *error) {
```

```
        UIAlertView * alert= [[UIAlertView alloc]initWithTitle:@"提示" message:
@"连接失败" delegate:nil cancelButtonTitle:@"好的" otherButtonTitles:nil, nil];
        [alert show];
    }];
}

@end
```

代码解析：

这里主要用到了 AFNetworking 的方法，将请求参数和接口地址传给 AFHTTPClient 实例，AFHTTPClient 实例异步请求或上传数据到服务端。如果成功，则将服务端返回数据(Json 格式)转换为 NSDictionary 并触发 Block。如果请求失败，则提示用户连接失败。

3. 编写注册事件代码

单击 RegisterViewController. m，重写 registChick：方法，代码如下：

```
- (IBAction)registChick:(id)sender {
    UIAlertView * alert= [[UIAlertView alloc]initWithTitle:@"提示" message:
nil delegate:nil cancelButtonTitle:@"好的" otherButtonTitles:nil, nil];
if(self.username.text.length < 1 || self.password.text.length < 1 || self.
passwordRepeat.text.length<1){
        alert.message=@"请完整输入";
        [alert show];
    }
    else if(self.username.text.length<6||self.username.text.length>12){
        alert.message=@"请输入 6~12 位用户名";
        [alert show];
    }
    else if(self.password.text.length<6||self.password.text.length>25){
        alert.message=@"请输入 6~25 位密码";
        [alert show];
    }
     else if (! [self.password.text isEqualToString: self.passwordRepeat.
text]){
        alert.message=@"两次密码输入不一致";
        [alert show];
    }
    else{
        NSDictionary * dic=@{@"act":@"register",@"username":_username.
text,@"password":_password.text};
```

```
        [AFHelper downDataWithDictionary:dic andBaseURLStr:@"http://localhost:
8080/meServer/" andPostPath:@"user.php?" success:^(NSDictionary *dic){
            UIAlertView * alert = [[UIAlertView alloc] initWithTitle:@"提示"
message:dic[@"msg"] delegate:nil cancelButtonTitle:@"好的" otherButtonTitles:
nil, nil];
            [alert show];
        }];
    }
}
```

代码解析：

程序首先对用户输入进行本地判断，如果不满足设定要求，则直接返回，否则将用户名和密码上传注册接口，完成注册。

上面代码中密码提交的是明文，真实开发中，密码需要先加密再提交到服务端，比如MD5加密。

4.5 用户注册的调试

测试用户注册功能前，请确保第2章的服务端配置成功，注册接口能够正常访问。

(1) 单击Xcode左上角的▶按钮，运行程序，如图4.56所示。

(2) 不输入任何内容，单击注册按钮，提示“请完整输入”，如图4.57所示。

(3) 输入账号“123”，输入密码“123”，重复输入密码“123”，提示“请输入6～12位用户名”，如图4.58所示。

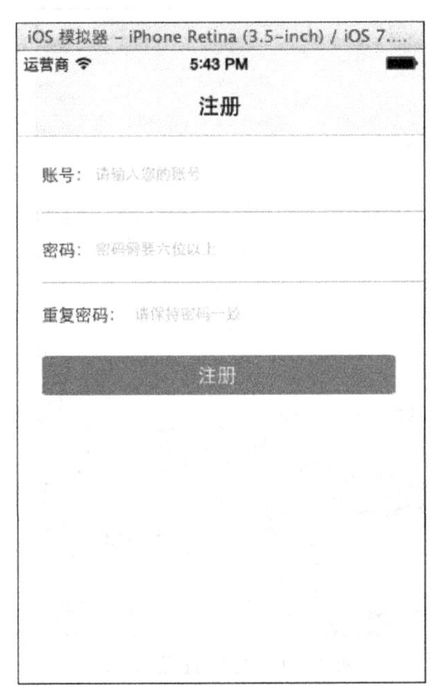

图4.56 注册界面

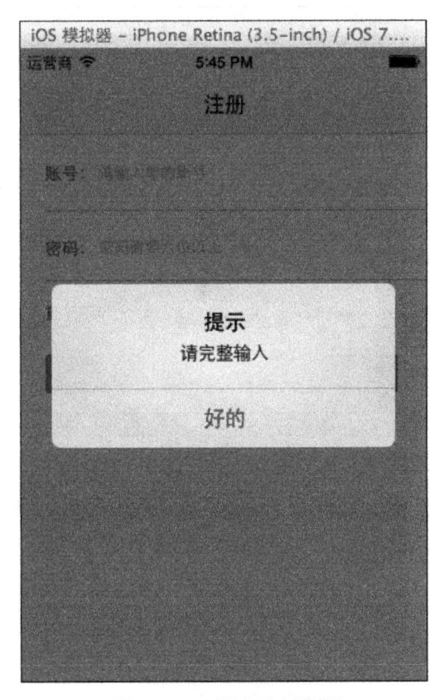

图4.57 提示对话框

(4) 输入账号“123456”,输入密码“123”,重复输入密码“123”,提示“请输入 6～25 位密码”,如图 4.59 所示。

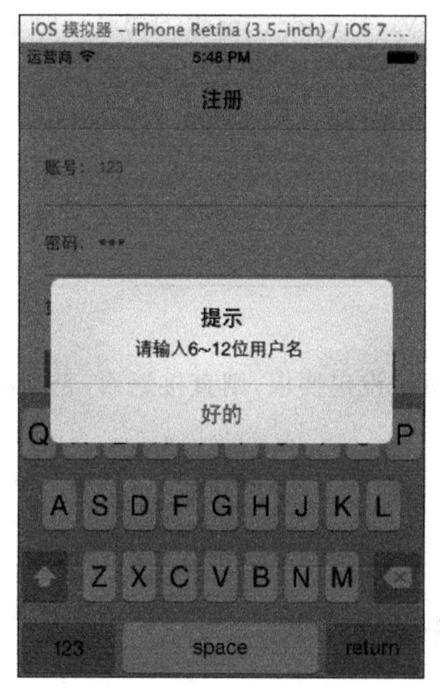

图 4.58　提示输入用户名

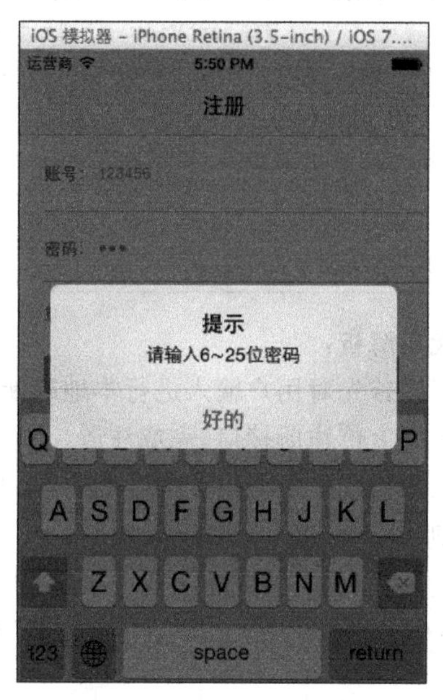

图 4.59　提示输入密码

(5) 输入账号“123456”,输入密码“123456”,重复输入密码“123”,提示“两次密码输入不一致”,如图 4.60 所示。

(6) 输入账号“123456”,输入密码“123456”,重复输入密码“123456”,提示“注册成功”,如图 4.61 所示。

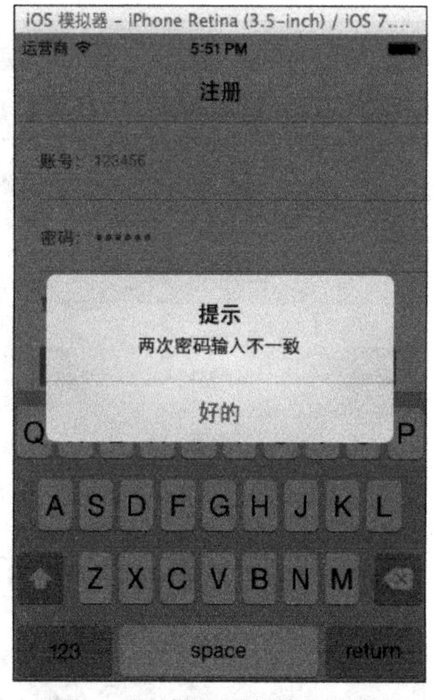

图 4.60　提示重新输入密码

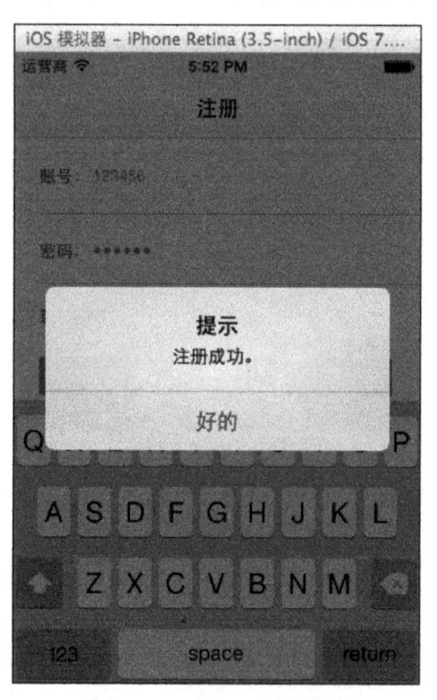

图 4.61　提示注册成功

(7) 打开数据库用户表,可以看到成功注册的账号,如图 4.62 所示。

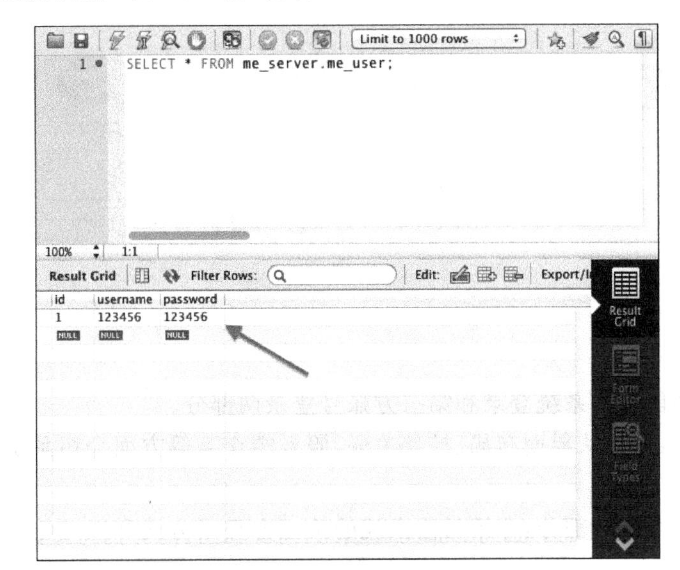

图 4.62 数据库写入成功

4.6 让用户免去注册的烦恼

对手机客户端来说,让用户注册、填写各种资料无疑提高了程序的使用门槛,但也会让用户觉得烦琐。可以通过第三方登录机制来有效降低程序的使用门槛,提高客户端的注册用户数量。第 5 章将具体介绍这种登录方式。

4.7 基础知识与技能回顾

本章主要介绍了如下内容:

(1) xib 布局界面。使用 xib 布局界面非常方便。首先操作非常简单,直接拖曳就能完成界面布局。其次,这种方式会自动生成代码,大大减少了开发者的代码编写量,提高了开发效率。

(2) UILabel、UIBtton、UITextField 等控件的使用。项目中,我们直接修改了控件的 Titile、位置等属性,通过属性面板能操作更多的控件属性。

(3) xib 上控件与 ViewController 类关联。通过连线方式关联 xib 上的控件,为其设置属性或事件,使开发者可以直接用代码操控控件,完成交互功能。

练 习

项目

功能描述:实现注册界面的功能,并注册一个用户名和密码都为"123456"的账号。

第5章 chapter 5

用户登录

用户登录功能分为系统登录和第三方账号登录两部分。

在系统登录部分，从界面布局、控件关联、服务端交互等方面介绍系统登录功能开发的完整过程。

在第三方账号登录部分，介绍目前主流的第三方登录方式和以 QQ 账号登录为案例的具体实现过程。

5.1 用户登录总体设计

用户登录时要求用户输入用户名、密码。用户单击登录按钮后，系统将用户账户和密码提交到后台。后台验证如果通过，则登录成功，否则登录失败。

第三方账号登录时要求用户输入第三方账号，如果验证通过，则登录成功，否则登录失败。

5.1.1 流程图

用户登录流程如图 5.1 所示。

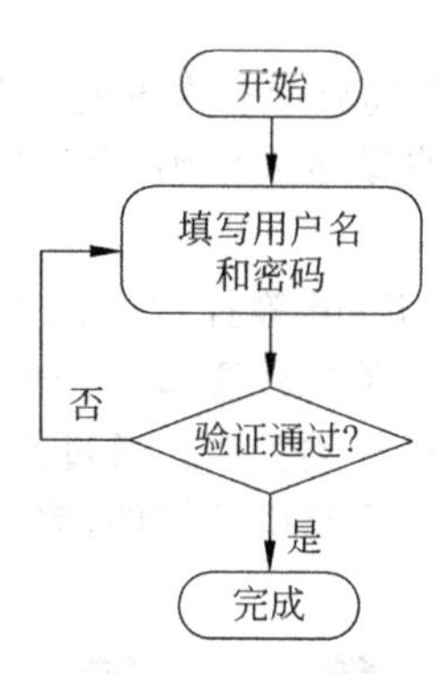

图 5.1 用户登录流程

5.1.2 时序图

用户登录的时序过程如图 5.2 所示。

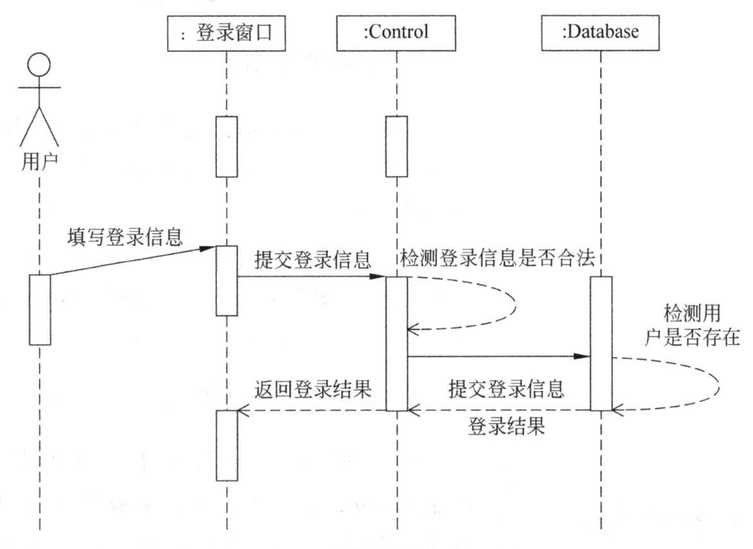

图 5.2 用户登录的时序过程

5.2 服务端接口的准备

本节将用到登录接口，下面给出其详细信息。

接口地址：http://localhost:8080/meServer/user.php?

调用方式：Post

返回数据格式：Json

登录接口请求参数及说明如表 5.1 所示。

登录接口返回字段及说明如表 5.2 所示。

表 5.1 登录接口请求参数

请求参数	必选	类 型	说 明
act	Y	string	login
username	Y	string	用户名
password	Y	string	密码

表 5.2 登录接口返回字段

返回字段	字段类型	字段说明
flag	string	0：失败，1：成功
msg	string	信息说明

5.3 用户登录的实现

本节首先介绍用 XIB 布局界面并与所属类进行代码关联。与纯代码布局相比，XIB 布局的最大好处是大量减少了代码量，降低了布局难度，方便初学者快速掌握。然后介绍客户端与服务端的交互，让服务端验证用户提交信息。

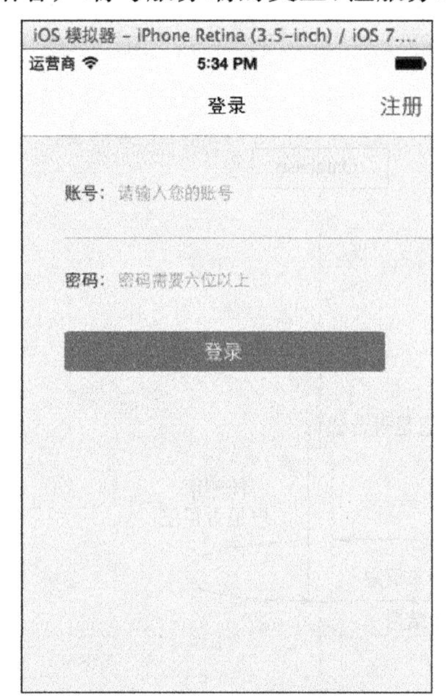

图 5.3 用户登录界面

用户登录界面如图 5.3 所示。

5.3.1 客户端代码开发

1. 创建登录页面

(1) 打开第 4 章已完成的项目，单击 Sections 文件夹，按下 command＋N，添加 Objective-C class，如图 5.4 所示。

(2) 在 Class 中输入 LoginViewController，在 Subclass of 中输入 FatherViewController，勾选 Also create XIB file，如图 5.5 所示。

2. 绘制登录页面界面

(1) 选择存储目录后，单击导航区下的 LoginViewController. xib，拖放两个 UILabel、两个 UITextField、一个 UIImageView、一个 UIButton 到

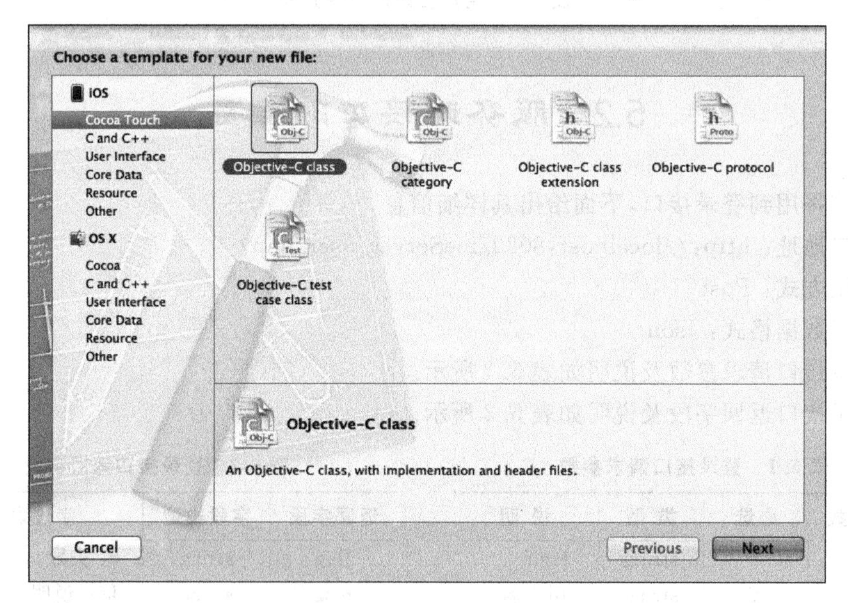

图 5.4 添加 Objective-C class

界面并调整位置，整体效果如图 5.6 所示。

(2) 设置两个 UILabel 属性，如图 5.7 所示。

(3) 设置两个 UITextField 属性，如图 5.8 所示。

(4) 设置 UIButton 属性，如图 5.9 所示。

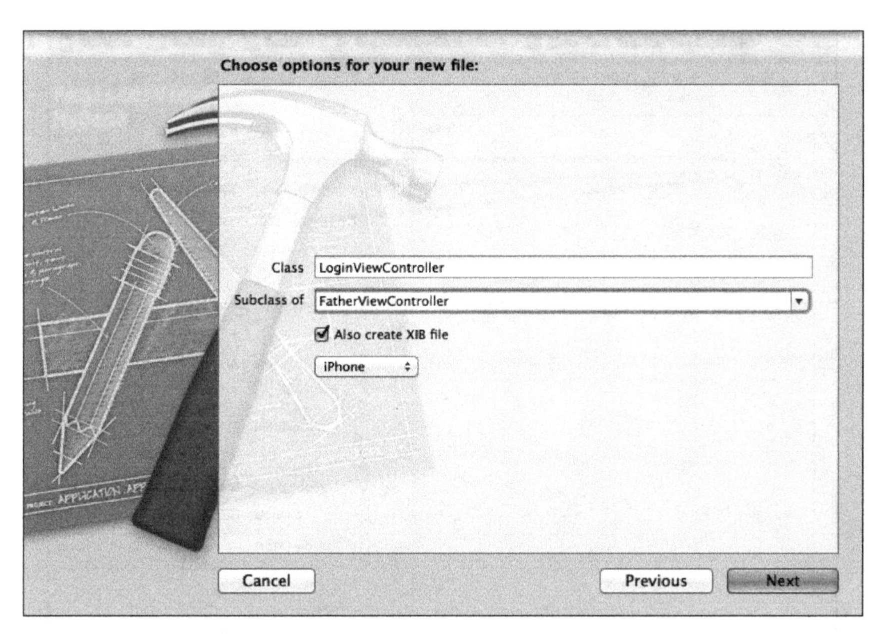

图 5.5 创建 LoginViewController

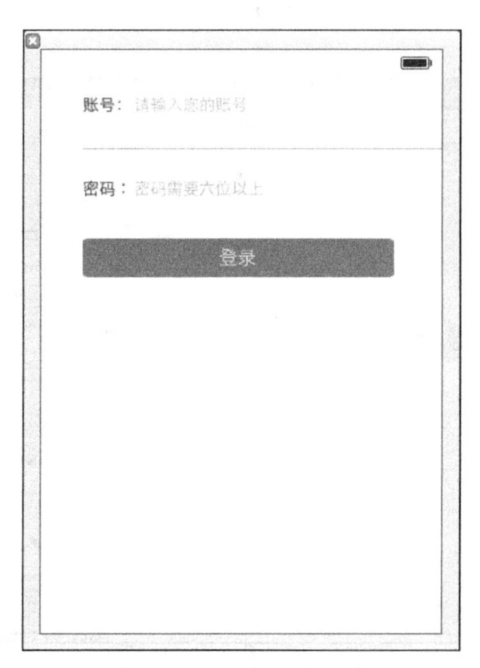

图 5.6 整体效果

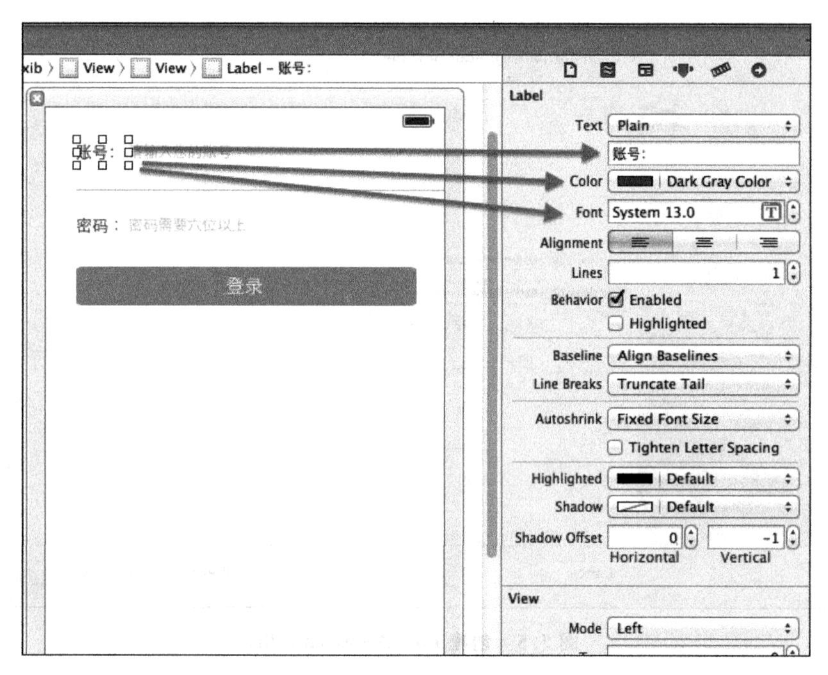

图 5.7 UILabel 属性

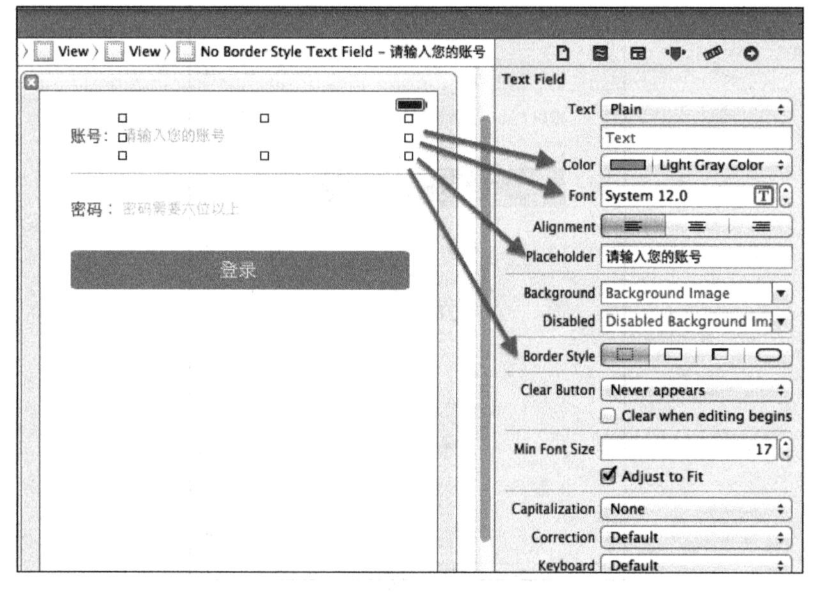

图 5.8 UITextField 属性

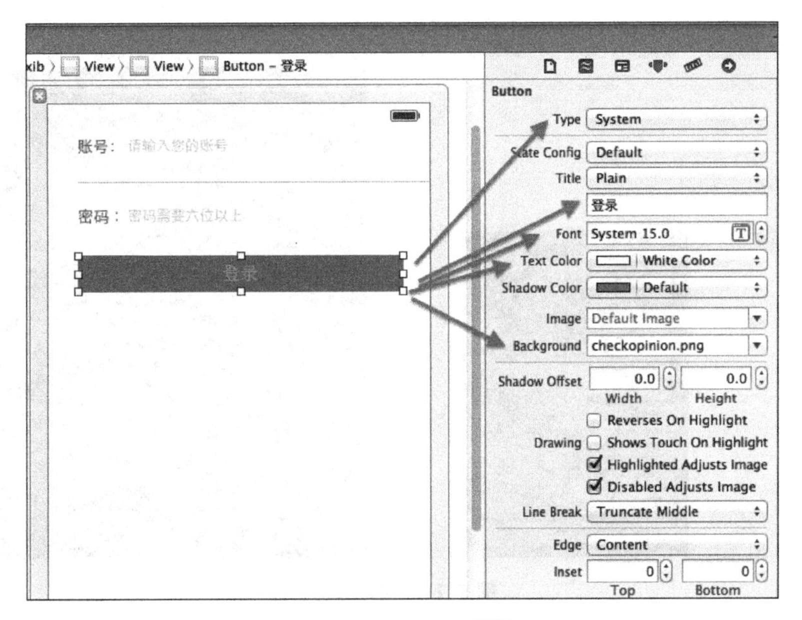

图 5.9 UIButton 属性

3. 关联控件

(1) 分别右击两个 UITextField 控件，关联 New Referencing OutLet 到 LoginView-Controller.h 中。将第一个控件命名为 username，第二个控件命名为 password，如图 5.10 所示。

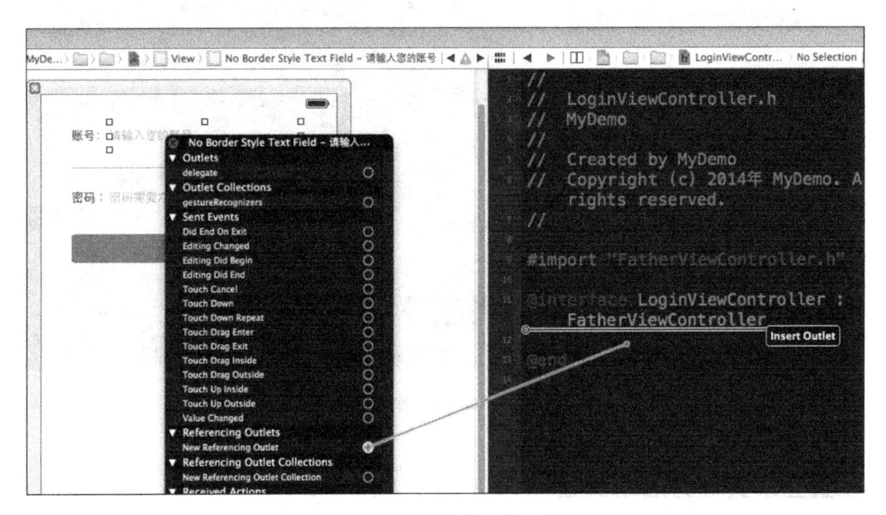

图 5.10 关联属性

(2) 右击 UIButton 控件，关联其 Touch Up Inside 事件到 LoginViewController.h 中，Name 为 registChick，如图 5.11 所示。

(3) 最终效果如图 5.12 所示。

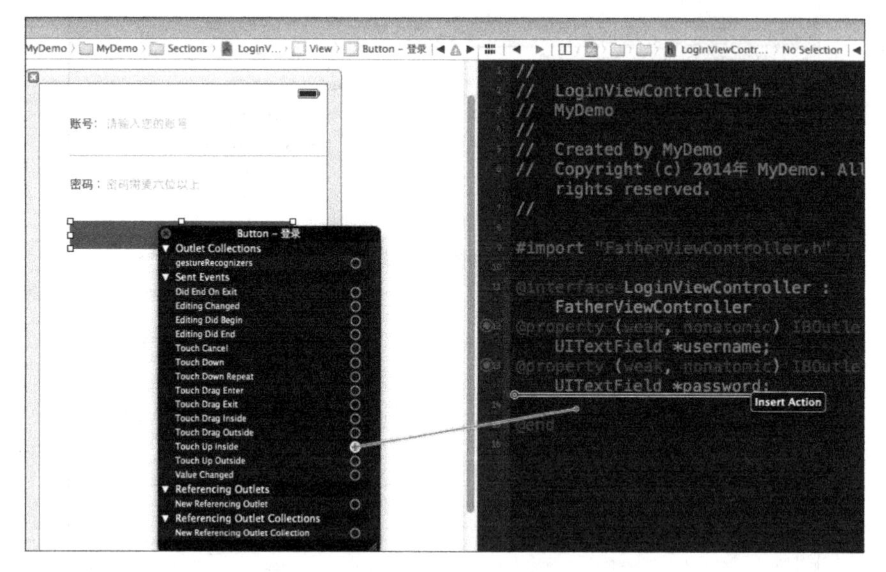

图 5.11 关联事件

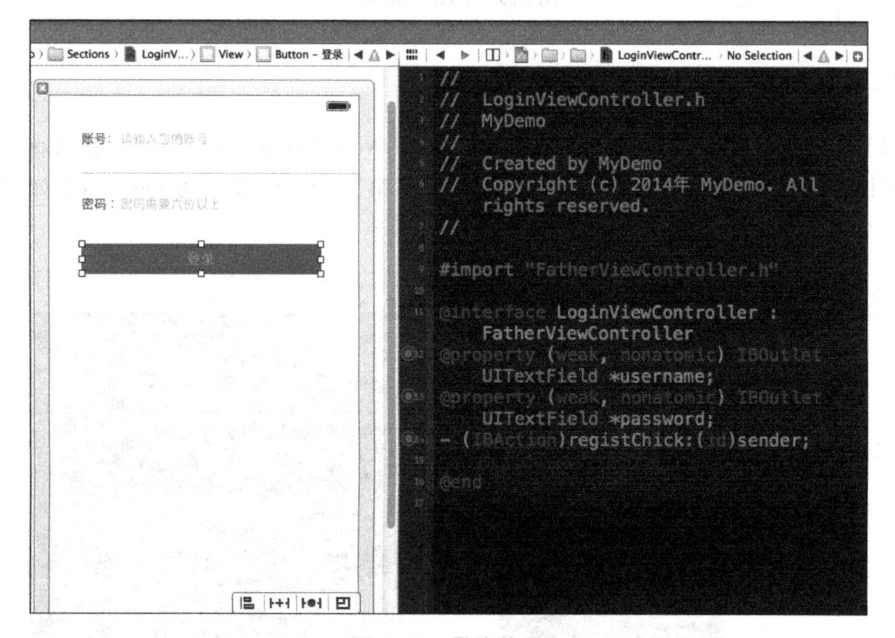

图 5.12 最终效果

4. 设置登录页面为启动页面

(1) 单击 AppDelegate.m，引入头文件 LoginViewController.h，并删除 RegisterViewController.h，如图 5.13 所示。

这里将登录页面设置为 window 的 rootViewController，因此不再需要 RegisterViewController.h，将其删除掉。

```
#import "AppDelegate.h"
#import "LoginViewController.h"
@implementation AppDelegate
```

图 5.13　引入头文件

（2）重写 application:didFinishLaunchingWithOptions:方法，代码如下：

```
- (BOOL)application: (UIApplication * ) application didFinishLaunchingWithOptions:
(NSDictionary * )launchOptions
{
    self.window = [[UIWindow alloc] initWithFrame: [[UIScreen mainScreen]
bounds]];
    LoginViewController * vc= [[LoginViewController alloc]initWithNibName:
@"LoginViewController" bundle:nil];
    UINavigationController  *  nav  =  [[ UINavigationController  alloc ]
initWithRootViewController:vc];
    self.window.rootViewController=nav;
    self.window.backgroundColor= [UIColor whiteColor];
    [self.window makeKeyAndVisible];
    return YES;
}
```

5. 连接登录页面与注册页面

（1）单击 LoginViewController. m，引入头文件 RegisterController. h，如图 5. 14 所示。

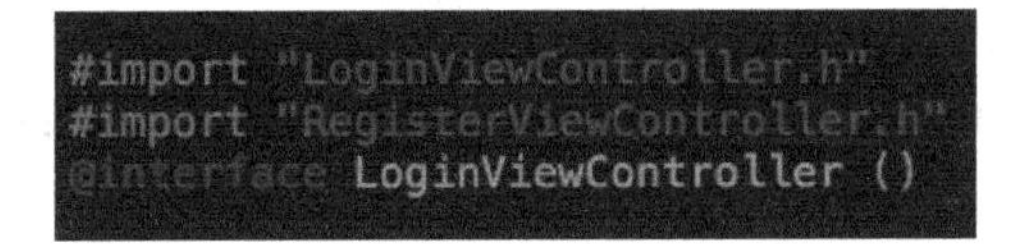

图 5.14　引入头文件

（2）重写 viewDidLoad 方法，添加代码如下：

```
- (void)viewDidLoad
{
    [super viewDidLoad];
    self.title=@"登录";
    UIBarButtonItem * rightBar= [[UIBarButtonItem alloc]initWithTitle:@"注
册"  style: UIBarButtonItemStyleBordered  target: self  action: @ selector
(goRegist)];
    self.navigationItem.rightBarButtonItem=rightBar;
```

```
}
- (void)goRegist{
    RegisterViewController * vc= [[RegisterViewController alloc] initWithNi-
bName:@"RegisterViewController" bundle:nil];
    [self.navigationController pushViewController:vc animated:YES];
}
```

代码解析：

在 viewDidLoad 方法中创建右上角的"注册"按钮，单击后跳转到注册页面。

pushViewController:animated 方法将视图控制器推入栈顶。

UINavigationController 通过栈来实现。添加一个 Controller 为入栈，释放一个 Controller 为出栈。

注意：

- 栈是先进后出。
- 栈顶是最后一个入栈的对象。
- 基栈是第一个入栈的对象（栈底）。

(3) 单击 RegisterViewController.m，找到 viewDidLoad 方法，添加代码如下：

```
- (void)viewDidLoad
{
    [super viewDidLoad];
    self.title=@"注册";
    [self createLeftBtnWithFrame: CGRectMake (0, 0, 24, 24) Image: [UIImage
imageNamed:@"backButtonNormal"]];
}
- (void)leftBarChick{
    [self.navigationController popViewControllerAnimated:YES];
}
```

代码解析：

重写父类 createLeftBtnWithFrame: Image: 方法，创建左上角的自定义按钮。单击按钮后回到登录页面。用户单击按钮后，NavigationController 进行出栈操作，弹出注册页面。

另外，NavigationController 还有两个出栈方法：

- popToRootViewControllerAnimated:方法：是否退到栈顶。
- popToViewController:animated:方法：退到指定的 viewController。

5.3.2 客户端与服务端交互

(1) 单击 LoginViewController.m，引入头文件 AFHelper，如图 5.15 所示。

```
#import "LoginViewController.h"
#import "RegisterViewController.h"
#import "AFHelper.h"
```

图 5.15 导入头文件

(2) 重写 loginChick:方法,代码如下:

```
- (IBAction)loginChick:(id)sender {
    if(self.username.text.length<1||self.password.text.length<1){
        UIAlertView * alert = [[UIAlertView alloc] initWithTitle:@"提示"
message:@"请完整输入" delegate:nil cancelButtonTitle:@"好的" otherButto-
nTitles:nil, nil];
        [alert show];
    }
    else{
        NSDictionary * dic=@{@"act":@"login",@"username":_username.text,
@"password":_password.text};
        [AFHelper downDataWithDictionary: dic andBaseURLStr: @" http:
//localhost: 8080/meServer/" andPostPath:@"user.php?" success:^(NSDictionary *
dic){
            if([dic[@"flag"]intValue]==1){
                NSLog(@"登录成功");
                [USER_DEFAULT setBool:YES forKey:@"isLogin"];
                [USER_DEFAULT setObject:_username.text forKey:@"username"];
                [USER_DEFAULT synchronize];
            }else{
                UIAlertView *alert=[[UIAlertView alloc]initWithTitle:@"提示"
message:dic[@"msg"] delegate:nil cancelButtonTitle:@"好的" otherButtonTitles:
nil, nil];
                [alert show];
            }
        }];
    }
}
```

代码解析:

上面代码中,程序先判断用户是否完整输入。如果没有完整输入,则提示"请完整输入",否则提交用户输入的用户名和密码给服务端验证。服务端如验证成功,则在 Debug 区打印"登录成功"。服务端若验证不成功,则弹出对话框,告知用户错误信息。

登录成功后,程序用 NSUserDefaults 来保存用户登录状态和用户名。

NSUserDefaults 进行用户轻量级的数据持久化,主要用于保存用户程序的配置等信息,以便下次启动应用程序后能恢复上次的设置。该数据实际上是以"键值对"形式保存的(类似于 NSDictionary),因此需要通过 key 来读取或者保存数据(value)。

具体使用如下：

获取一个 NSUserDefaults 引用，代码如下：

```
NSUserDefaults * userDefaults= [NSUserDefaults standardUserDefaults];
```

保存数据，代码如下：

```
[userDefaults setObject:@"tom" forKey:@"username"];
[userDefaults synchronize];
```

userDefault 不是立即写入数据，而是根据时间戳定时把缓冲中的数据写入本地磁盘。因此在调用 setObject:forKey 方法写入数据后，应该调用 synchornize 方法强制写入数据，以免数据遗失。

读取数据，代码如下：

```
NSString * username= [userDefaults objectForKey:@"username"];
```

NSUserDefaults 可以存储的数据类型包括 NSData、NSString、NSNumber、NSDate、NSArray、NSDictionary。如果需要保存其他类型，如 UIImage，应该进行编码(即 archive)，或者将它转换为 NSData、NSNumber 或 NSString。

5.4 用户登录的调试

测试用户登录功能前，请确保第 2 章服务端配置成功，登录接口能够正常访问。

第 4 章中已经注册了一个用户名、密码都为 123456 的账号。如果没有注册，请返回第 4 章重新注册或在浏览器输入以下网址注册。

```
http://localhost:8080/meServer/user.php?act=register&username=123456&password=123456
```

(1) 在顶部工具栏找到 Scheme 按钮。单击该按钮，选择下拉菜单中的 iPhone Retina(3.5-inch)，如图 5.16 所示。

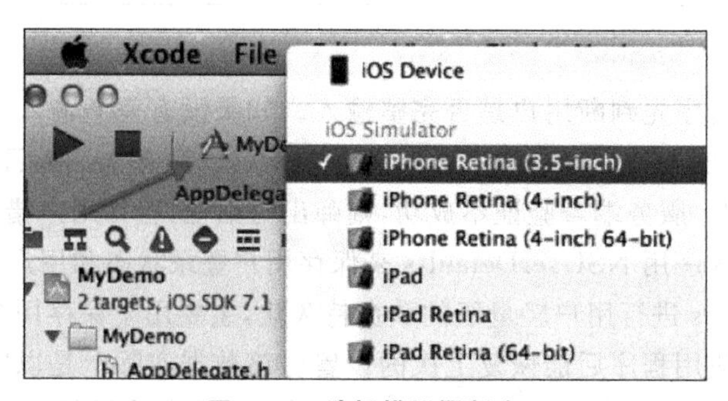

图 5.16　选择模拟器版本

(2) 单击 Xcode 左上角的▶按钮，运行程序，如图 5.17 所示。

(3) 单击“登录”按钮，提示“请完整输入”，如图 5.18 所示。

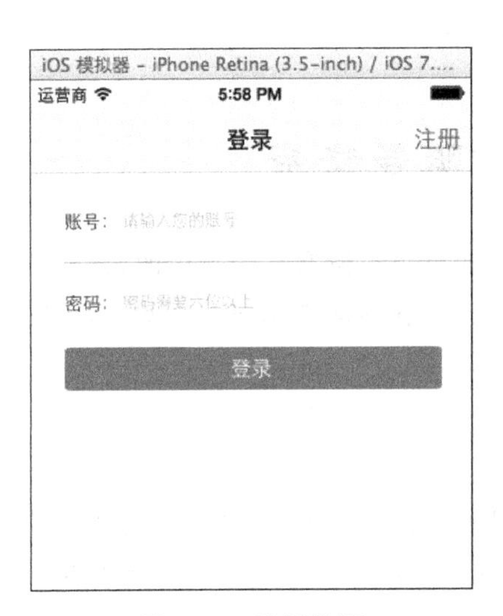

图 5.17 登录界面

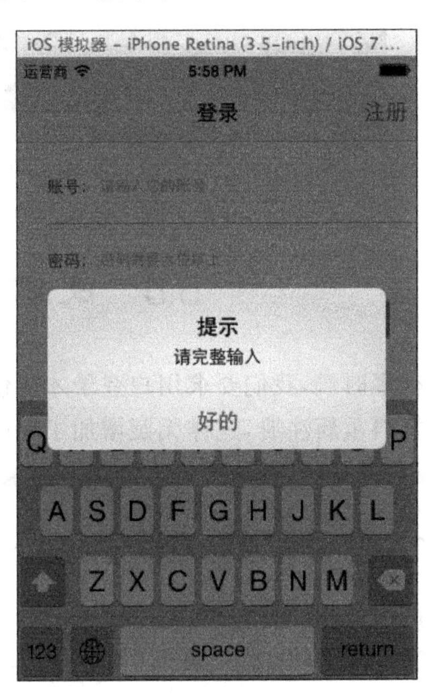

图 5.18 提示“请完整输入”

(4) 输入账号 123456、密码 111111。单击“登录”按钮，提示“用户账号或密码错误”，如图 5.19 所示。

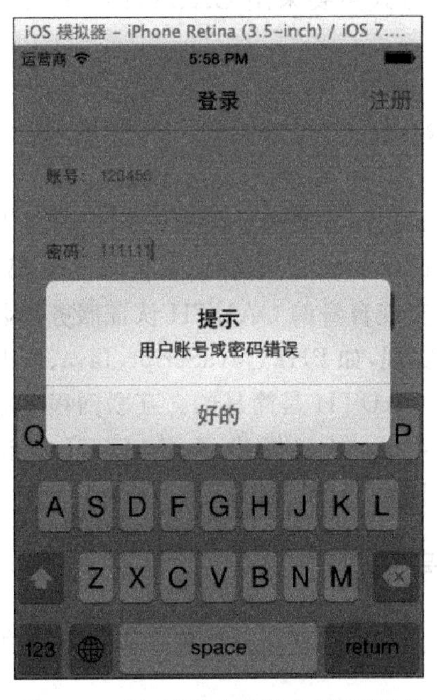

图 5.19 提示“用户账号或密码错误”

(5) 输入账号 123456、密码 123456，单击“登录”按钮，登录页面下移消失，在 Debug 区打印“登录成功”，如图 5.20 所示。

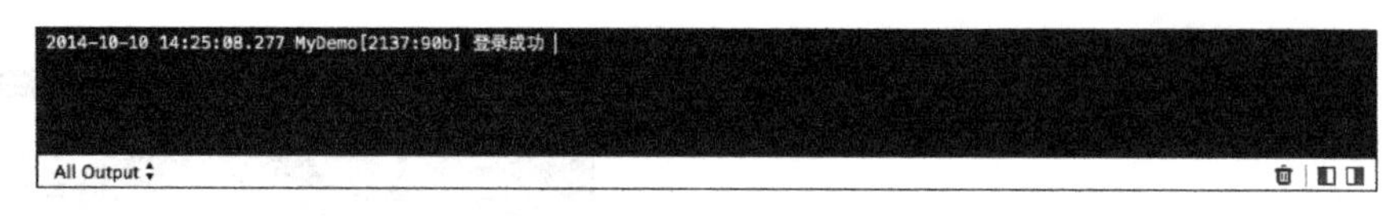

图 5.20 提示“登录成功”

5.5 使用第三方账号登录

很多时候，我们要求用户登录才能使用程序，如果用户没有我们程序的账号，就只能填写资料重新注册，这样无疑增加了程序的使用门槛，容易流失用户。因此，为了简化登录过程、降低注册门槛，第三方账号登录应运而生。

5.5.1 什么是第三方账号

就本章的登录程序而言，用户为第一方，登录程序为第二方，有自己会员体系的平台称为第三方，比如 QQ、微博。而用户在这些平台的账号，相对于我们的程序就叫第三方账号。

5.5.2 第三方账号登录方式

目前，第三方账号登录方式主要采用 OAUTH 技术，OAUTH 认证授权分为四个步骤：

(1) 获取未授权的 Request Token。

(2) 获取用户授权的 Request Token。

(3) 用授权的 Request Token 换取 Access Token。

(4) 使用 Access Token 访问或修改受保护资源。OAUTH 协议为用户资源的授权提供了一个安全、开放而又简易的标准。同时，任何第三方都可以使用 OAUTH 认证服务，任何服务提供商都可以实现自身的 OAUTH 认证服务，因而 OAUTH 是开放的。业界提供了 OAUTH 的多种实现，如 PHP、JavaScript、Java、Ruby 等各种语言开发包，大大节约了程序员的时间，因而 OAUTH 是简易的。互联网很多服务如 Open API，很多大公司如 Google、Yahoo、Microsoft 等都提供了 OAUTH 认证服务，这些都足以说明 OAUTH 标准已逐渐成为开放资源授权的标准。

5.5.3 使用第三方账号登录

本节以 QQ 平台为例，介绍如何在项目里集成第三方账户登录功能。

1. 导入 QQSDK

(1) 到 http://wiki.connect.qq.com/sdk%E4%B8%8B%E8%BD%BD#iOSSDK 下载最新版本的 QQ 登录 iOS SDK,如图 5.21 所示。

(2) 下载后的 iOS SDK 目录结构如图 5.22 所示。

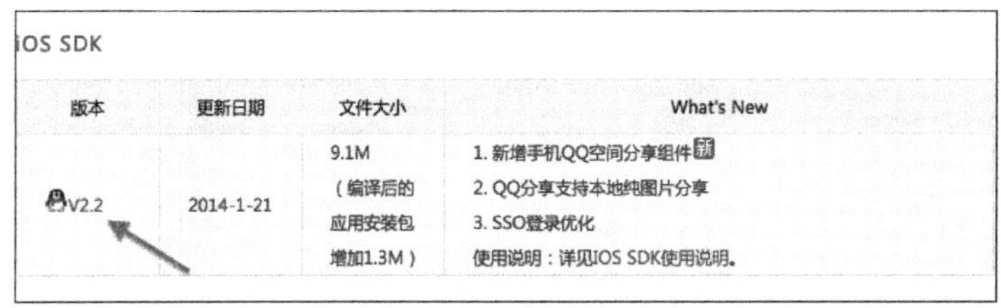

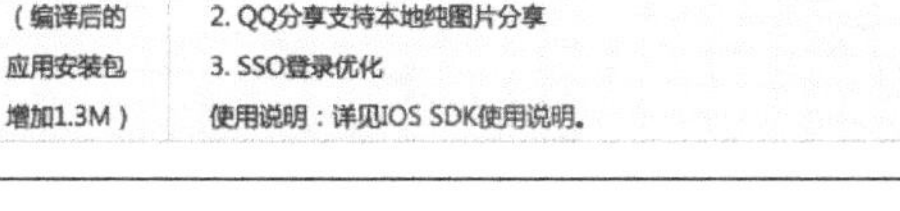

图 5.21 下载 iOS SDK

图 5.22 iOS SDK 目录结构

iOS SDK 文件夹中带有两个文件:

TencentOpenAPI.framework——打包了 iOS SDK 的头文件定义和具体实现。

TencentOpenApi_iOS_Bundle.bundle——打包了 iOS SDK 需要的资源文件。

(3) 将 iOS SDK 文件夹中的 TencentOpenAPI.framework 和 TencentOpenApi_IOS_Bundle.bundle 文件拖放到项目 Venders 文件夹中,如图 5.23 所示。

图 5.23 拖放 iOS SDK 文件夹中的文件

(4) 在弹出的对话框中选中 Create groups for any added folders,勾选 Copy items into destination group's folder(if needed),在 Add to targets 中选择 MyDemo 之后单击 Finish,将 iOS SDK 的 framework 文件加入到开发工程中,如图 5.24 所示。

(5) 由于 QQ SDK 使用了系统的一些框架,因此需要将这些框架添加到工程中。单击工程名,选择 Build Phases,接着单击 Linked Binary and Libraries(4 items),最后单击"+"图标,如图 5.25 所示。

(6) 在搜索框中输入 Security.framework,双击搜索结果中的 Security.framework,

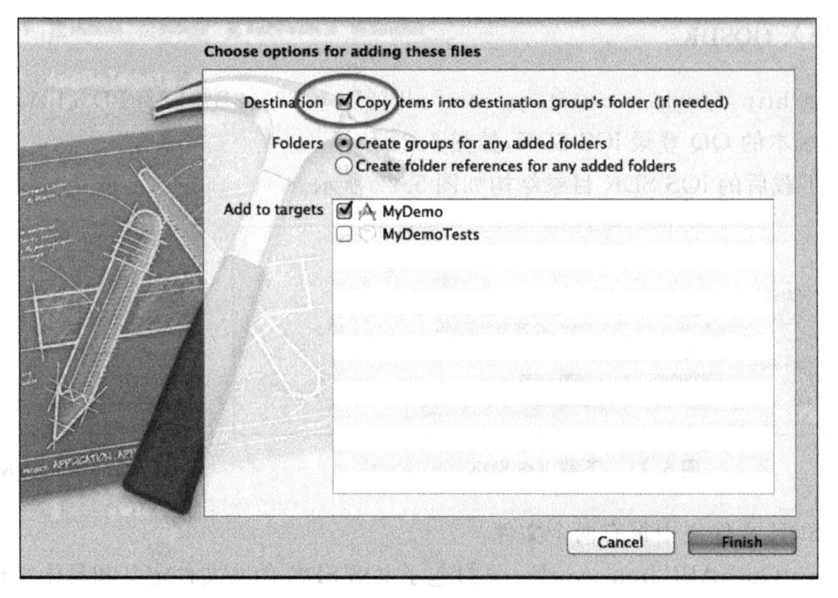

图 5.24　将框架加入到工程中

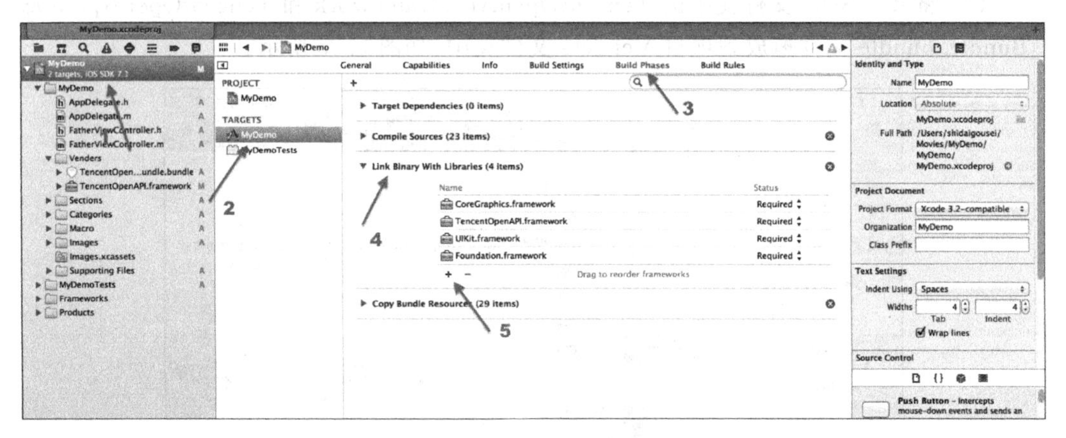

图 5.25　添加所需框架

完成依赖库的添加，如图 5.26 所示。

(7) 依次添加下面的依赖库：

```
libiconv.dylib
SystemConfiguration.framework
CoreGraphics.Framework
libsqlite3.dylib
CoreTelephony.framework
libstdc++.dylib
libz.dylib
```

添加完成后如图 5.27 所示。

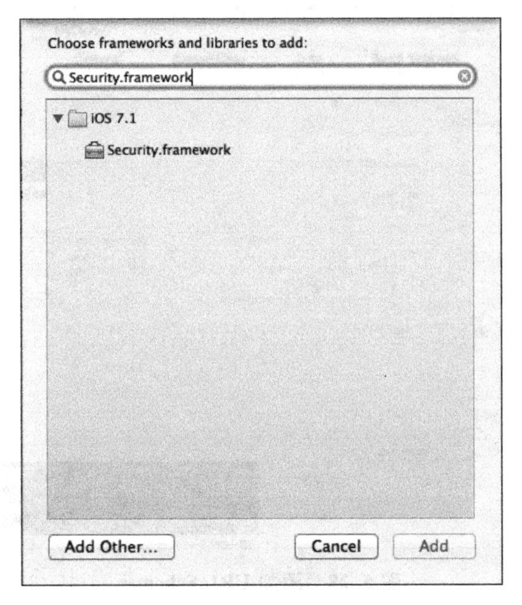

图 5.26 搜索框架

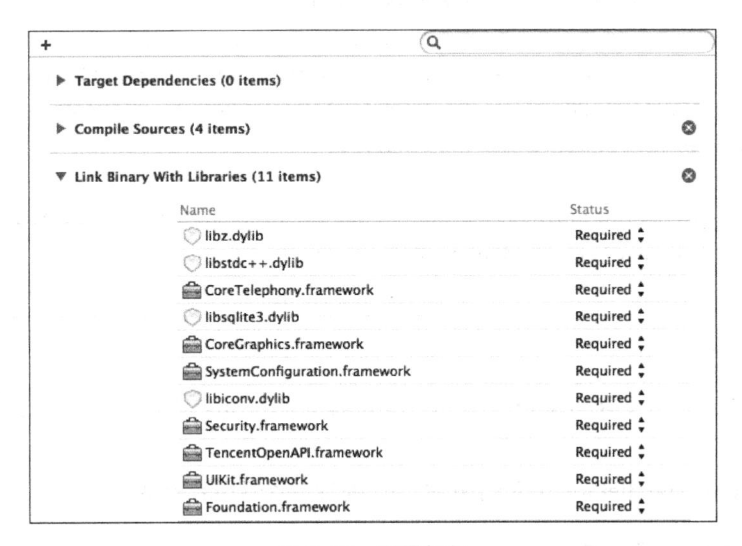

图 5.27 所需框架

(8) 在 Info 标签栏的 URL Types 添加一条新的 URL Schemes，新的 Schemes＝tencent＋appid，这里使用 tencent222222，如图 5.28 所示。

在 iOS 里，程序之间都是相互隔离，目前并没有一个有效的方式来做程序间通信，然后 iOS 程序可以很方便地注册自己的 URL Scheme，这样就可以通过打开特定 URL 的方式来传递参数给另外一个程序。

tencent222222 是 QQ SDK 提供的示例 Schemes，真实开发中，开发者需要到 QQ 开放平台注册账号，申请成为开发者并建立应用，拿到对应的唯一标示。这里为便于讲解，我们使用 QQ SDK 提供的测试 Schemes。

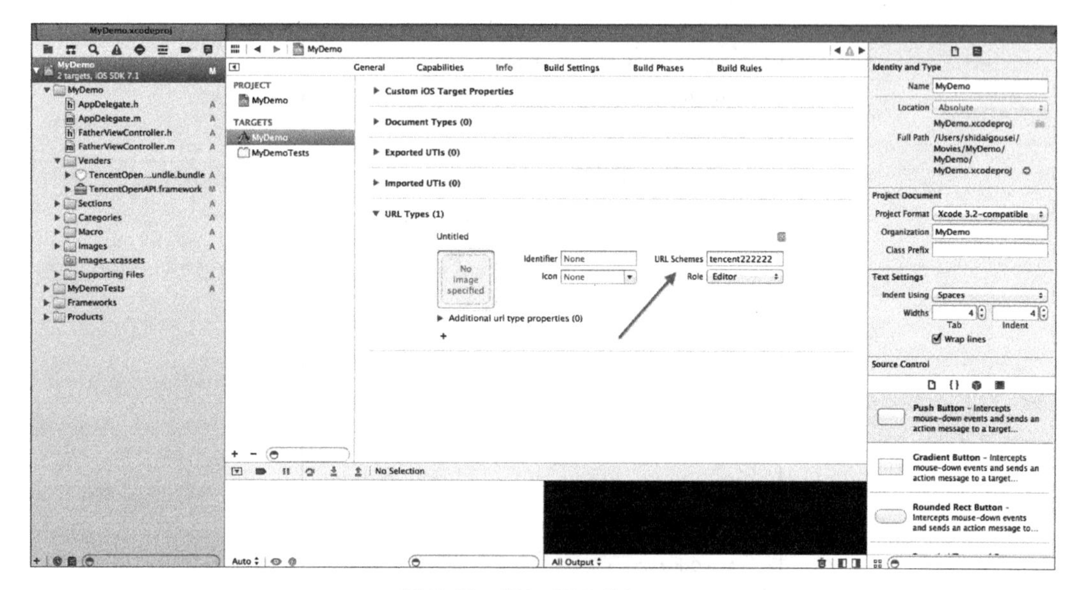

图 5.28　添加 URL Schemes

(9) 单击 AppDelegate.m 文件，添加代码如下：

```
- (BOOL) application: (UIApplication * ) application openURL: (NSURL * ) url
sourceApplication:(NSString *)sourceApplication annotation:(id)annotation{
return [TencentOAuth HandleOpenURL:url];
}
-(BOOL)application:(UIApplication *)application handleOpenURL:(NSURL *)url{
return [TencentOAuth HandleOpenURL:url];
}
```

代码解析：

application:openURL:sourceApplication:annotation:方法——请求委托打开一个 URL 资源。

application: handleOpenURL:方法——处理一个 URL 资源。

2. 添加 QQ 登录入口

(1) 打开 LoginViewController.xib 文件，拖放一个 UIButton 到界面上，双击命名为“QQ 登录”，如图 5.29 所示。

(2) 为 UIButton 添加 Touch Up Inside 事件，命名为 loginQQ，如图 5.30 所示。

3. 编写“QQ 登录”按钮单击事件

在编写单击事件代码之前，先介绍一个 iOS 开发中非常重要的知识点：protocol(协议)和 delegate(委托)。

protocol：声明一组方法，列出其参数和返回值，自身不实现这些方法，而是让使用它

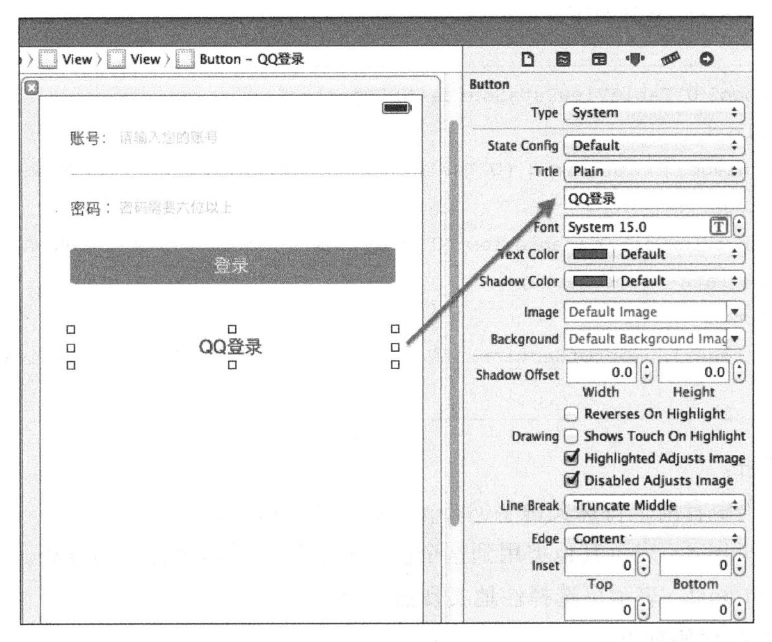

图 5.29　添加“QQ 登录”按钮

图 5.30　关联事件

的类来实现。

delegate：顾名思义，就是委托别人。

下面以 UITableView 为例，介绍 protocol 和 delegate 的用法。

新建 protocol,代码如下：

```
@protocol UITableViewDataSource<NSObject>
@required
- (NSInteger) tableView: (UITableView * ) tableView numberOfRowsInSection:
(NSInteger)section;
- (UITableViewCell * )tableView:(UITableView * )tableView cellForRowAtIndexPath:
(NSIndexPath * )indexPath;
@optional
- (NSInteger)numberOfSectionsInTableView:(UITableView * )tableView;
@end
```

代码解析：

上面代码中有两个特殊关键字@required 和@optional。

- @required　表示其他类用到这个协议时必须实现这个协议的方法。
- @optional　表示可选择性地实现这些方法。

使用协议,代码如下：

```
@interfacetestViewController: UIViewController<UITableViewDataSource>
@property (strong, nonatomic)UITableView * table;
@end
@implementationtestViewController
- (void)viewDidLoad{
    //其他代码
    _table.dataSource=self ;
}
- (NSInteger) tableView: (UITableView * ) tableView numberOfRowsInSection:
(NSInteger)section{
}
- (UITableViewCell * )tableView:(UITableView * )tableView cellForRowAtIndexPath:
(NSIndexPath * )indexPath{
}
@end
```

代码解析：

要使用 UITableViewDataSource 协议,需要先设置 UITableView 的 delegate 属性,并实现 tableView：numberOfRowsInSection：和 tableView：cellForRowAtIndexPath：方法。

下面结合项目介绍如何使用 protocol 和 delegate。

(1) 单击 LoginViewController.m 文件,导入头文件,添加 TencentSessionDelegate 协议,并添加 TencentOAuth 属性,如图 5.31 所示。

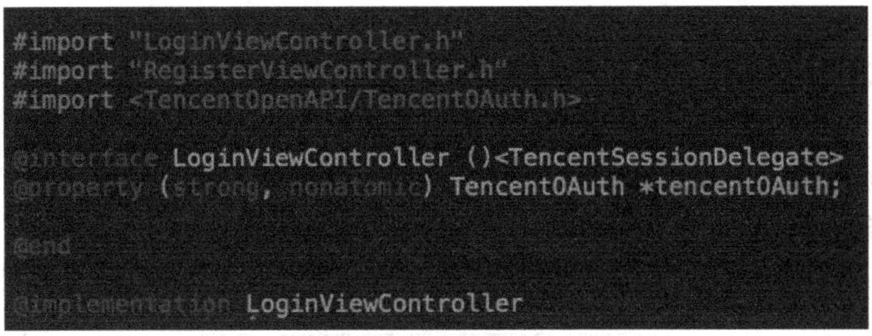

```
#import "LoginViewController.h"
#import "RegisterViewController.h"
#import <TencentOpenAPI/TencentOAuth.h>

@interface LoginViewController ()<TencentSessionDelegate>
@property (strong, nonatomic) TencentOAuth *tencentOAuth;

@end

@implementation LoginViewController
```

图 5.31 添加代码

(2) 重写 loginQQ 方法，并实现 TencentSessionDelegate 协议方法，代码如下：

```
-(IBAction)loginQQ:(id)sender {
    _tencentOAuth=[[TencentOAuth alloc] initWithAppId:@"222222" andDelegate:
    self];
    [_tencentOAuth authorize:@[@"get_user_info"] inSafari:NO];
}
-(void)tencentDidLogin
{
    if (_tencentOAuth.accessToken && 0 !=[_tencentOAuth.accessToken length])
    {
        NSLog(@"QQ账号登录成功");
        NSLog(@"%@",_tencentOAuth.accessToken);
    }
    else
    {
        NSLog(@"登录不成功 没有获取 accesstoken");
    }
}
-(void)tencentDidNotLogin:(BOOL)cancelled
{
    if (cancelled)
    {
        NSLog(@"用户取消登录");
    }
    else
    {
        NSLog(@"登录失败");
    }
}
-(void)tencentDidNotNetWork
{
    NSLog(@"无网络连接,请设置网络");
}
```

代码解析：

单击“QQ 登录”按钮，TencentOAuth 将创建一个实例，同时将 AppId 上传，验证用户的合法性。

tencentDidLogin、tencentDidNotLogin：、tencentDidNotNetWork 都为协议方法。开发者可以修改这些方法中的代码，实现相应的功能。

(3) 单击左上角的▶按钮运行程序。单击“QQ 登录”按钮后，界面如图 5.32 所示。

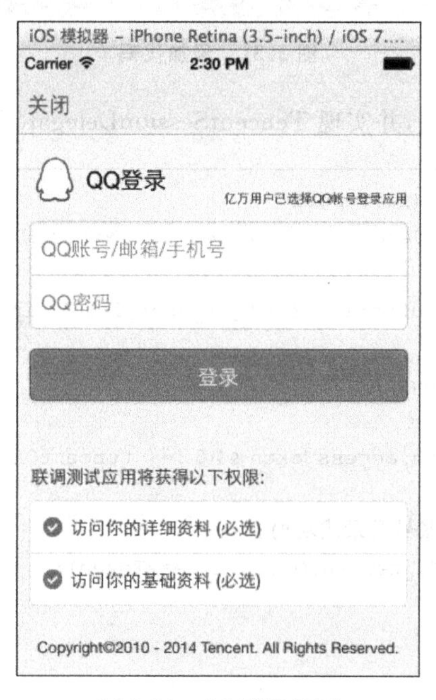

图 5.32 QQ 登录界面

(4) 正确输入 QQ 账号、密码，单击登录，Xcode 提示“QQ 账户登录成功”，如图 5.33 所示。如提示的是其他内容，请按照其内容检查错误。

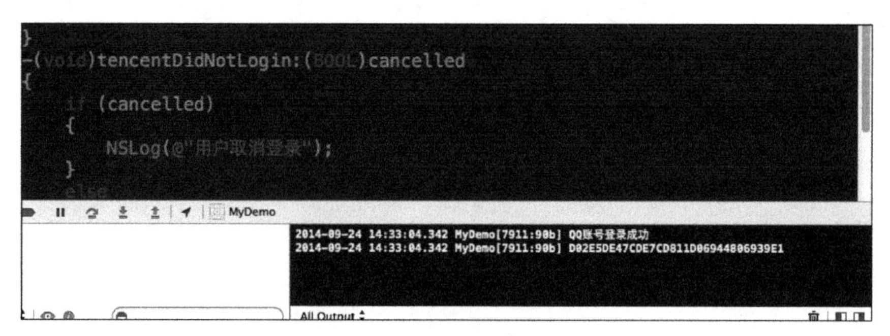

图 5.33 QQ 登录提示

5.6 基础知识与技能回顾

本章主要介绍了系统登录功能和第三方账号登录的使用方法。使用第三方账号的好处如下：

(1) 对用户而言，第三方账号登录带来的最大好处就是方便。一方面可以省去注册流程；另一方面也不用费心去记各种账号、密码，可以用微博等账号一号走遍天下。此外，通过微博、QQ等平台更容易与好友分享并实现互动。这些好处让许多比较“懒”的用户更倾向于用第三方账号登录。

(2) 对开发者而言，第三方账号登录的好处主要是有以下三点：

- 提升用户的注册转化率，降低进入的门槛。
- 利用微博、QQ等平台的资源，提高自己的知名度。
- 省去自主登录体系的开发工作。

练　　习

项目1

功能描述：实现第4章已注册用户的登录。

项目2

功能描述：使用新浪微博账号登录。

第 6 章 chapter 6

向用户展示内容

在手机上展示内容，具有很多局限性。例如，手机屏幕小影响页面不能显示太多内容；环境光线的变化影响页面设计不能过于花哨；流量限制影响不能有太多的图片和样式。

如何调整内容展示方式，使内容在小屏幕的手机上也能更友好地展示？本章将介绍目前手机端的主流做法。

6.1 数据库的准备

第 2 章已介绍了如何搭建项目服务端并配置完成本项目所需的后台数据。下面简要介绍数据库的准备。

商户表用来存储商户信息，如图 6.1 所示。

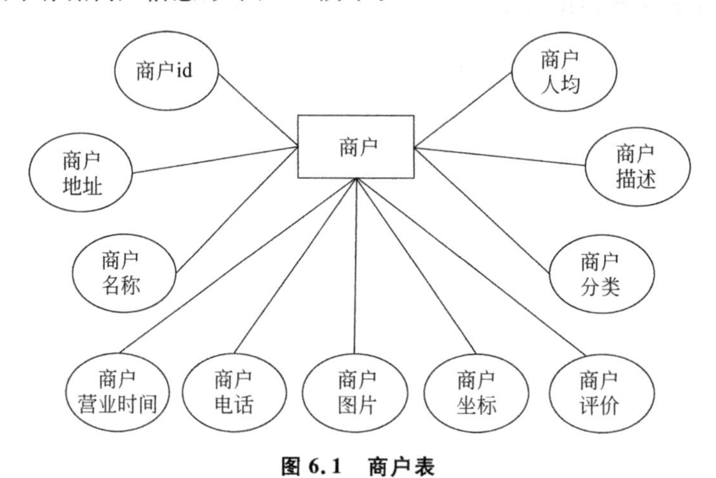

图 6.1 商户表

SQL 语句如下：

```
CREATE TABLE 'me_shop' (
  'shop_id' int(11) unsigned NOT NULL AUTO_INCREMENT,
  'shop_address' varchar(120) NOT NULL,
  'shop_name' varchar(80) NOT NULL,
```

```
  'shop_time' varchar(30) NOT NULL,
  'shop_phone' varchar(20) NOT NULL,
  'shop_image' varchar(120) NOT NULL,
  'shop_longitude' varchar(16) NOT NULL COMMENT '经度',
  'shop_latitude' varchar(16) NOT NULL COMMENT '纬度',
  'shop_type' varchar(2) NOT NULL,
  'shop_desc' varchar(255) NOT NULL,
  'shop_spend' varchar(12) NOT NULL,
  'shop_create_time' timestamp NOT NULL DEFAULT CURRENT_TIMESTAMP ON UPDATE
CURRENT_TIMESTAMP,
  PRIMARY KEY ('shop_id')
) ENGINE=InnoDB AUTO_INCREMENT=3 DEFAULT CHARSET=utf8;
```

6.2 服务端接口的准备

本节将用到商户信息接口，下面给出接口的详细信息。

接口地址：http://localhost:8080/meServer/shop.php?

调用方式：Post

返回数据格式：Json

商户信息接口请求参数及说明如表 6.1 所示。

表 6.1 商户信息接口请求参数

请求参数	必选	类　型	说　　明	请求参数	必选	类　型	说　　明
act	Y	string	getShops	latitude	N	string	纬度
pageSize	Y	string	每页数量	longitude	N	string	经度
page	Y	string	页号				

商户信息接口返回字段及说明如表 6.2 所示。

表 6.2 商户信息接口返回字段

返回字段	字段类型	字 段 说 明	返回字段	字段类型	字 段 说 明
flag	string	0：失败，1：成功	shopLongitude	string	商户经度
msg	string	信息说明	shopName	string	商户名字
vendorList	JsonArray	商户信息数组	shopPhone	string	商户电话
shopAddress	string	地址	shopSpend	string	商户人均花费
shopDesc	string	描述	shopTime	string	商户营业时间
shopId	string	商户 id	shopType	string	商户类型
shopImage	string	商户图片	distance	string	商户距离
shopLatitude	string	商户纬度			

6.3 实现内容展示静态页面

本项目采用列表页+详情页的方式来展示商户内容。用户通过列表页浏览商户的图片、名称等简要信息,如图 6.2 所示。

详情页面展示商户的更多内容。单击详情页面左上方的后退按钮可回到列表页,如图 6.3 所示。

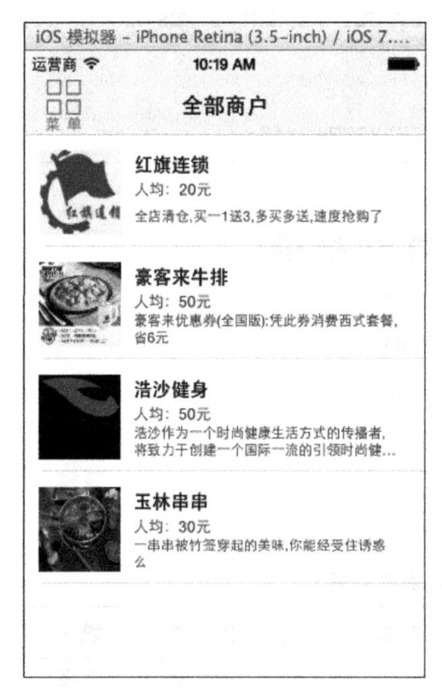

图 6.2　商户列表

图 6.3　商户详情

6.3.1 图文列表展示

在第 4 章,已经导入了侧滑框架。下面介绍如何使用侧滑框架。首先需要添加侧滑框架的左侧滑菜单页,左侧滑菜单页主要是各个功能的入口。

1. 添加左侧滑菜单页

(1) 选中 Sections 文件夹,按下 command+N,选择 iOS 下的 Objective-C class,如图 6.4 所示。

(2) 在 Class 中填写 LeftViewController,在 Subclass of 中填写 UIViewController,勾选 Also create XIB file,使用 XIB 来进行视图布局,如图 6.5 所示。

(3) 选择保存路径后,单击 LeftViewController. xib,将 xib 背景色设置为 Dark Gray Color,如图 6.6 所示。

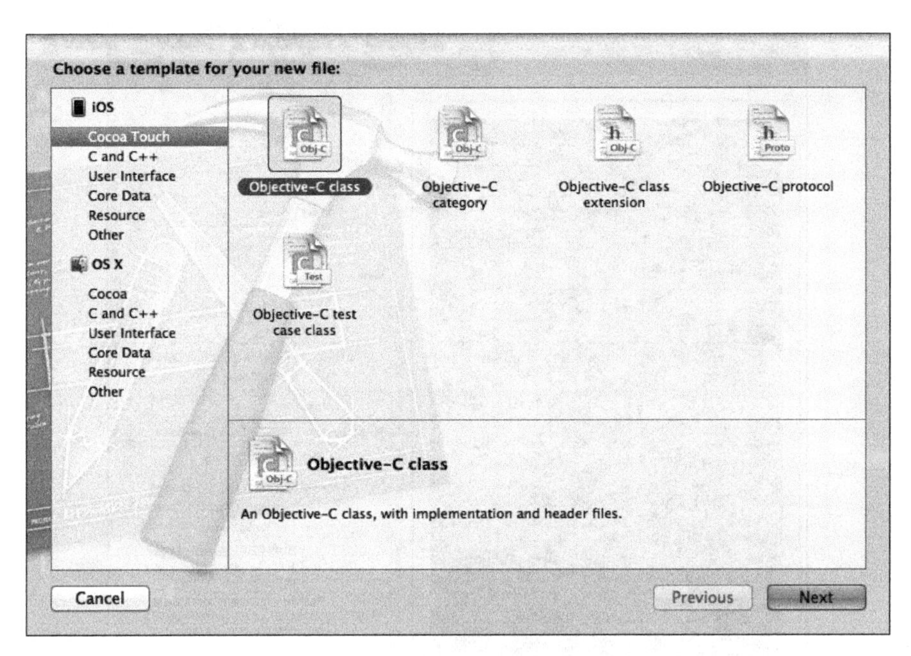

图 6.4 选择 Objective-C class

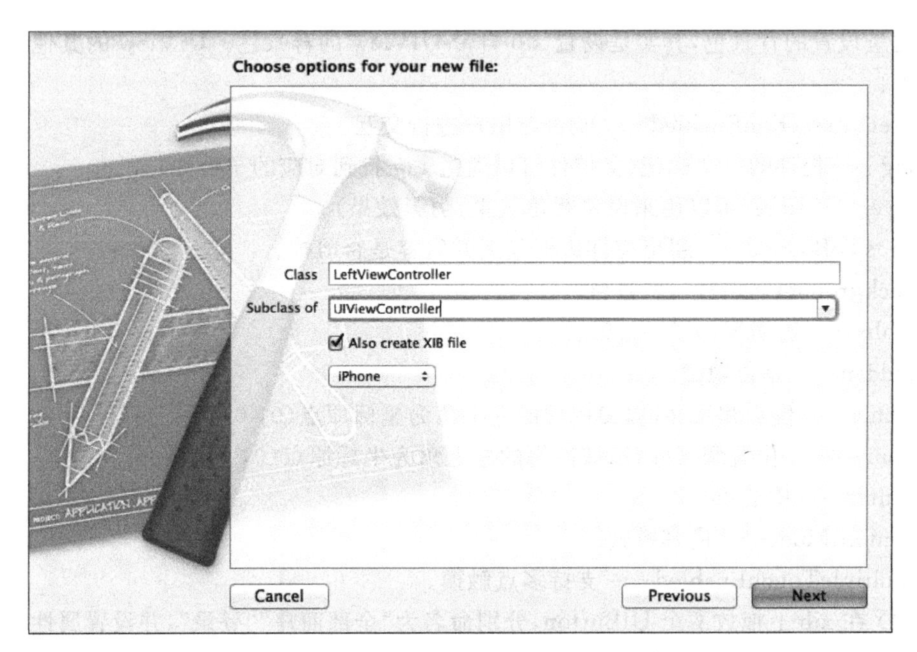

图 6.5 填写信息

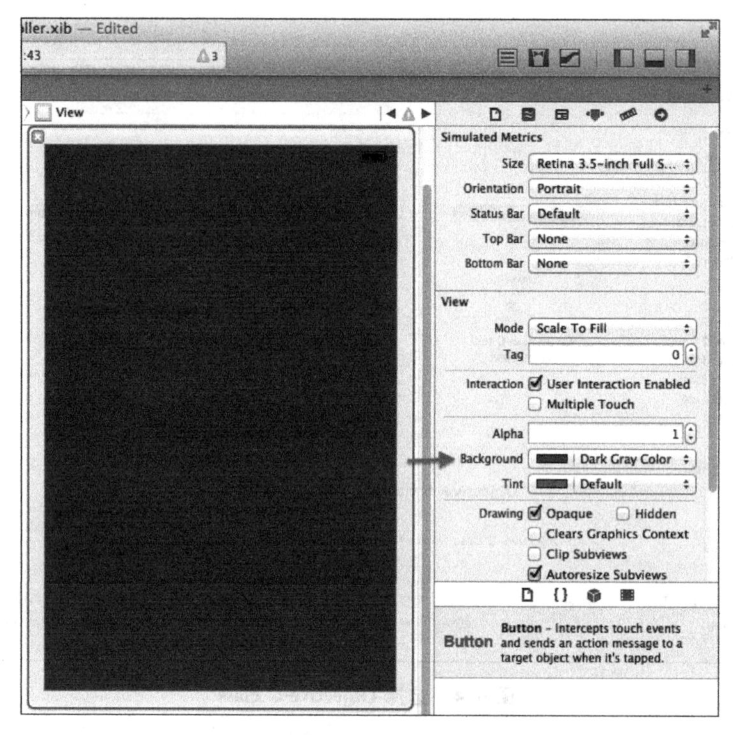

图 6.6 设置 xib 背景颜色

这里设置的背景色，其实是设置 xib 自带 UIView 的背景色。UIView 的其他常用属性如下：

userInteractionEnabled——能否与用户进行交互。

tag——控件的一个标记(父控件可以通过 tag 找到对应的子控件)。

layer——图层(可以用来设置圆角效果、阴影效果)。

clipsToBounds——超出控件边框范围的内容是否剪掉。

backgroundColor——背景色。

alpha——透明度(0.0～1.0)。

hidden——是否隐藏。

frame——位置和尺寸(以父控件的左上角为坐标原点(0，0))。

bounds——位置和尺寸(以该控件的左上角为坐标原点(0，0))。

center——中心点。

contentMode——内容模式。

multipleTouchEnabled——支持多点触摸。

(4) 在 xib 上拖放两个 UIButton，分别命名为“全部商户”“登录”，并设置属性。另设置“登录”按钮 Tag 的属性为 100，如图 6.7 所示。

将“登录”按钮 tag 值设置为 100，父视图就能根据 tag 值找到“登录”按钮。需要注意的是，同一父视图上控件的 tag 值不能相同。根据 tag 值获取“登录”按钮的方法如下：

```
UIButton * button= (UIButton * ) [self.view viewWithTag:100];
```

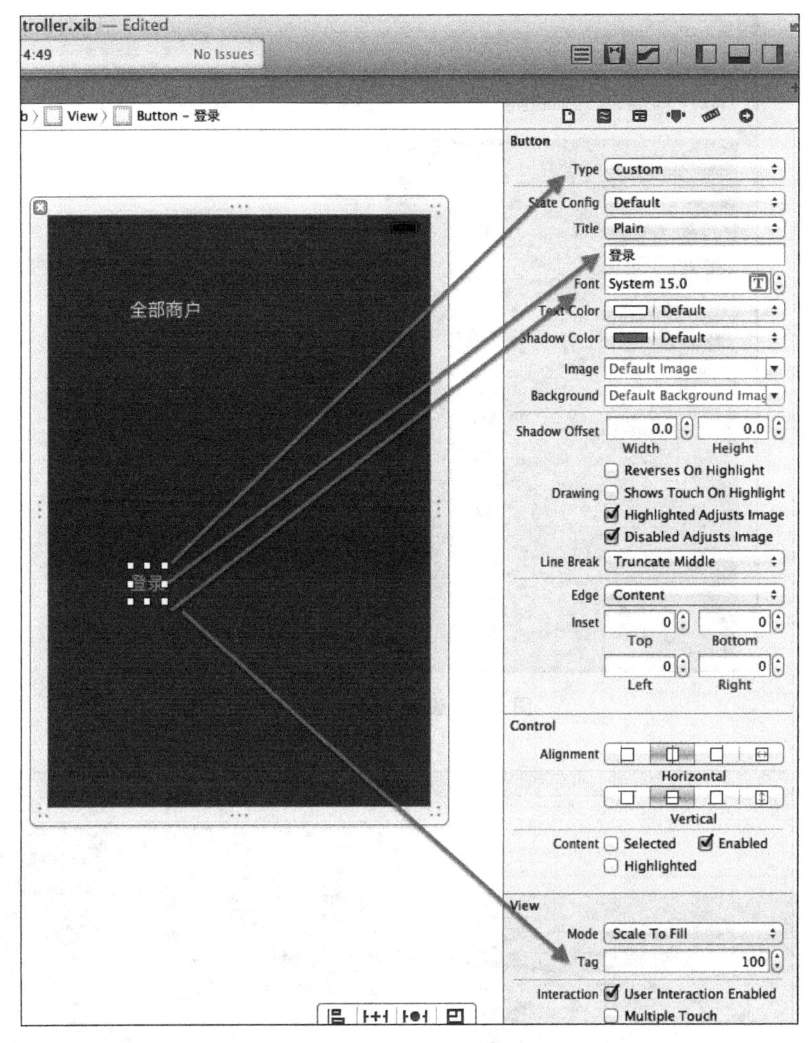

图 6.7 设置按钮属性

由于 viewWithTag:方法返回的是一个 UIView，因此需要转换为 UIButton。

(5) 在 xib 上拖放两个 UIImageView，第 1 个 UIImageView 的 Image 属性为 iconMechantNormal. png，第 2 个 Image 的属性为 userIcon. png，如图 6.8 所示。

(6) 单击 Xcode 右上角的按钮，切换助手编辑页到 LeftViewController. h，为“全部商户”和“登录”按钮分别关联 Touch Up Inside 事件，如图 6.9 所示。

(7) 将“全部商户”按钮 Touch Up Inside 事件命名为 goAllShop，“登录”按钮的 Touch Up Inside 事件命名为 goLogin，如图 6.10 所示。

此外，还需要一个页面作为侧滑框架的中心页面，下面添加全部商户页，作为侧滑中心页。

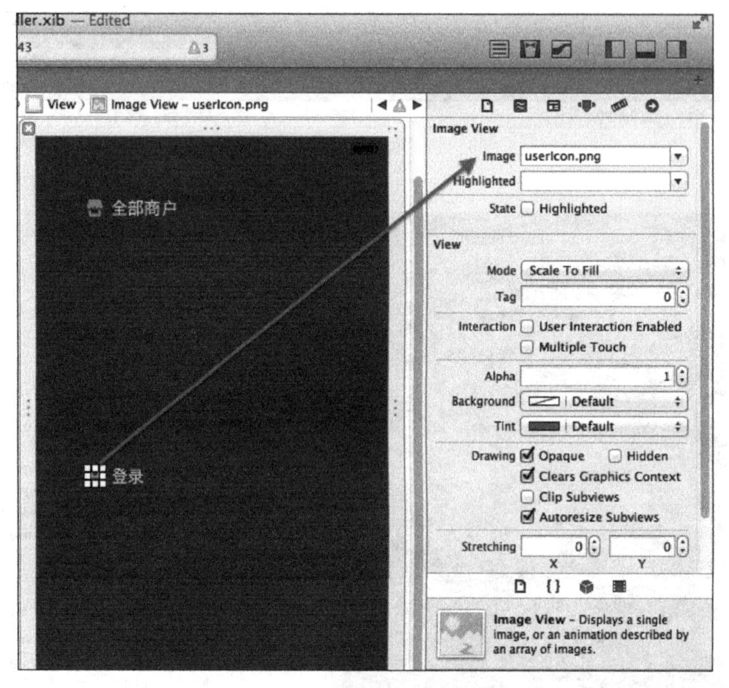

图 6.8　设置 Image 属性

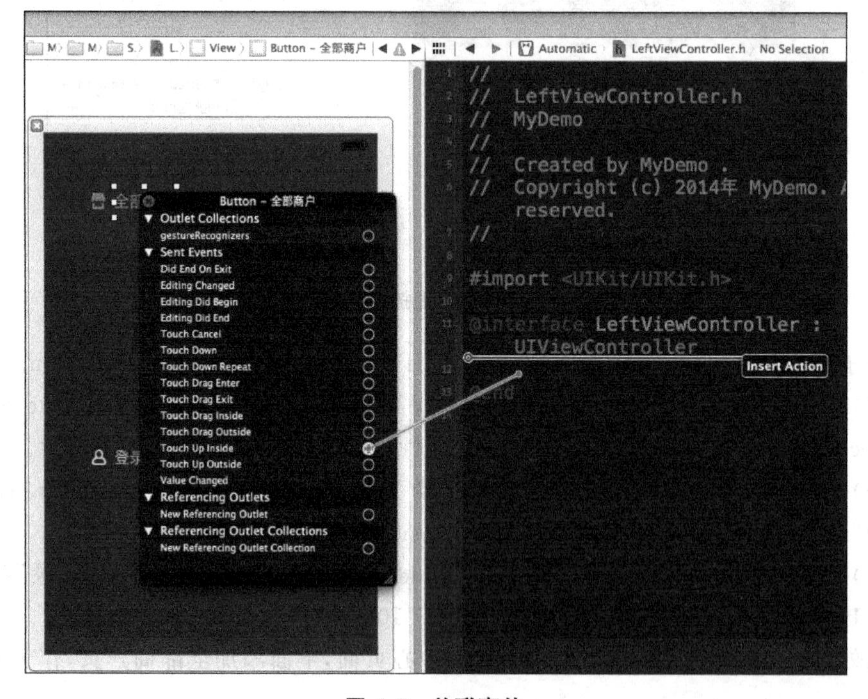

图 6.9　关联事件

```
#import <UIKit/UIKit.h>

@interface LeftViewController :
    UIViewController
- (IBAction)goAllShop:(id)sender;
- (IBAction)goLogin:(id)sender;

@end
```

图 6.10 方法命名

2. 添加全部商户页面

(1) 选中 Sections 文件夹,按下 command+N,选择 iOS 下的 Objective-C class,单击 Next,如图 6.11 所示。

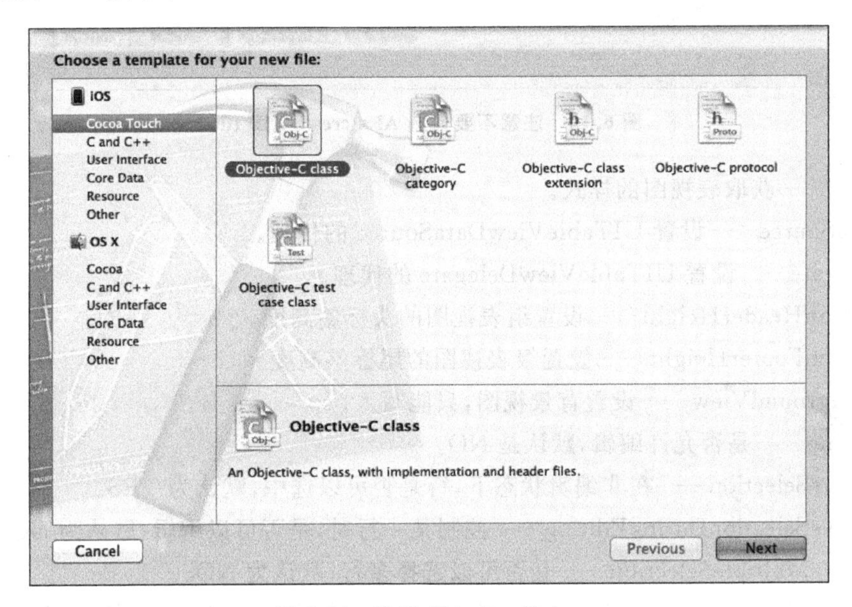

图 6.11 选择 Objective-C class

(2) 在 Class 中填写 AllShopViewController,在 Subclass of 中填写 FatherView Controller,此处不要勾选 Also create XIB file,单击 Next,如图 6.12 所示。

3. 编写 AllShopViewController.m 代码

在全部商户页面,需要用到 iOS 开发中的一个非常重要的控件:UITableView。UITableView 的功能很强大,多用于数据的显示,通常需要实现 UITableView 的两个协议:UITableViewDataSource、UITableViewDelegate。

UITableView 常用的属性如下:

frame——设置控件的位置和大小。

backgroundColor——设置控件的颜色。

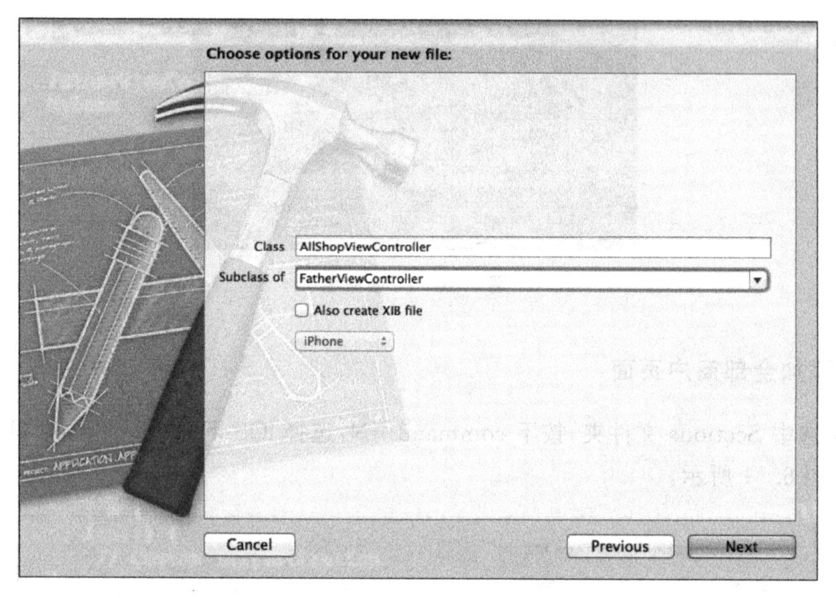

图 6.12 注意不要勾选 Also create XIB file

style——获取表视图的样式。

dataSource——设置 UITableViewDataSource 的代理。

delegate——设置 UITableViewDelegate 的代理。

sectionHeaderHeight——设置组表视图的头标签高度。

sectionFooterHeight——设置级表视图的尾标签高度。

backgroundView——设置背景视图，只能写入。

editing——是否允许编辑，默认是 NO。

allowsSelection——在非编辑状态下，行是否可以选中，默认为 YES。

allowsSelectionDuringEditing——控制某一行时，是否可以编辑，默认为 NO。

allowsMultipleSelection——是否可以选择多行，默认为 NO。

allowsMutableSelectionDuringEditing——在选择多行的情况下，是否可以编辑，默认为 NO。

sectionIndexMinimumDisplayRowCount——显示某个组索引列表在右边，当行数达到这个值时，默认是 NSInteger 的最大值。

sectionIndexColor——选择某个部分的某行改变这一行上文本的颜色。

sectionIndexTrackingBackgroundColor——设置选中某个部分的背景颜色。

separatorStyle——设置单元格分隔线的样式。

separatorColor——设置选中单元格分隔线的颜色。

tableHeaderView——设置组表的头标签视图。

tableFooterView——设置组表的尾标签视图。

UITableView 常用的代理方法如下：

tableView:heightForRowAtIndexPath:方法——设置每行的高度。

tableView:tableView heightForHeaderInSection:方法——设置组表的头标签高度。
tableView:tableView heightForFooterInSection:方法——设置组表的尾标签高度。
tableView:viewForHeaderInSection:方法——自定义组表的头标签视图。
tableView:viewForFooterInSection:方法——自定义组表的尾标签视图。
tableView:didSelectRowAtIndexPath:方法——获取选中的单元格的响应事件。
tableView:didDeselectRowAtIndexPath:方法——获取未选中的单元格的响应事件。
下面结合本项目介绍 UITableView 的具体用法。
(1) 添加 UITableView 协议和 UITableView 属性,代码如下:

```
@interface AllShopViewController ()<UITableViewDataSource,UITableViewDelegate>
@property (nonatomic,strong)UITableView * table;
@end
```

代码解析:

由于后面将把 UITableView 的 dataSource 属性和 delegate 属性委托给当前视图控制器,因此需添加 UITableViewDataSource 和 UITableViewDelegate 协议。

(2) 添加左上角按钮和 UITableView,代码如下:

```
- (void)loadView{
    [super loadView];
    self.title=@"全部商户";
    [self createLeftMenuBtn];
    _table= [[UITableView alloc]initWithFrame:CGRectMake(0, 0, App_SCREEN_
WIDTH, App_SCREEN_HEIGHT-64) style:UITableViewStylePlain];
    [self.view addSubview:_table];
    _table.rowHeight=86;
    _table.dataSource=self ;
    _table.delegate=self;
}
```

代码解析:

[self createLeftMenuBtn]是 Father 类的一个方法,这个方法用来创建页面左上角的按钮,实现侧滑的开、关。由于父类已经实现单击方法,因此这里不用再重复实现。

(3) 实现 UITableView 协议方法,代码如下:

```
- (NSInteger) tableView: (UITableView * ) tableView numberOfRowsInSection:
(NSInteger)section{
    return5;    //由于没有对接服务端,暂时返回 5 行
}
- (UITableViewCell * )tableView:(UITableView * )tableView cellForRowAtIndexPath:
(NSIndexPath * )indexPath{
    //指定 cellIdentifier
```

```
    static NSString * cellIdentifier=@"shopCell";
    UITableViewCell  * cell= [tableView dequeueReusableCellWithIdentifier:
cellIdentifier];
    if (cell==nil) {
      cell= [ [UITableViewCell  alloc] initWithStyle: UITableViewCellStyleDefault
reuseIdentifier:cellIdentifier];
    }
    return cell;
}
```

代码解析：

这里要介绍 UITableView 非常重要的功能：UITableViewCell 复用。

UITableView 属于 lazy loading，也就是只加载会在界面上显示的部分。举个例子，UITableview 的高度是 460，每个 UITableViewCell 的高度是 230，这样，手机界面最多显示 2 个 UITableViewCell。当向上划动时，第 1 个 UITableViewCell 部分离开界面，第 3 个 Cell 部分出现在界面。这时，UITableView 会再创建第 3 个 UITableViewCell。继续上滑，当第 2 个 UITableViewCell 离开界面，第 4 个 UITableViewCell 出现时，UITableView 不会创建第 4 个 UITableViewCell，而是直接复用第 1 个 UITableViewCell。也就是说，无论 UITableView 里有多少行数据，只创建 3 个 UITableViewCell 来展示这些数据。

在 tableView：cellForRowAtIndexPath：方法中，程序首先定义一个静态字符串：shopCell；接着程序调用 UITableView 的 dequeueReusableCellWithIdentifier：方法查找是否有这个静态字符串标示的 UITableViewCell，如果有，则直接返回这个 cell；如果没有，则创建一个 UITableViewCell，并用这个静态字符串来标示。这就是 UITableViewCell 的复用过程。

4. 创建自定义 UITableViewCell

列表页用到了 UITableView，而 UITableView 的内容主要由 UITableViewCell 组成。UITableView 中显示的每一个单元都是一个 UITableViewCell 对象，其初始化函数 initWithStyle：reuseIdentifier：比较特别，与一般 UIView 的初始化函数不同。这主要是为了效率考虑，因为在 tableView 快速滑动的过程中，频繁地调用 alloc 对象是比较费时的，于是引入了 cell 的重用机制。

UITableViewCell 的常用属性有：

accessoryType——右边辅助按钮的样式。

backgroundView——cell 的背景。

detailTextLabel——cell 的 contentview 中 detail 的文字内容。

editingAccessoryType——cell 进入编辑模式时的辅助按钮样式。

selectedBackgroundView——cell 被选中时的背景。

selectionStyle——cell 的选中状态样式。

textLabel——cell 的 contentview 中的 textlabel 文字内容。

虽然系统提供了几种 UITableViewCell 的样式，但比较简单，不适合项目需要，通常开发者需要自定义 UITableViewCell 来显示内容。自定义 UITableViewCell 可以用纯代码和 xib 来创建。

下面介绍如何使用 xib 创建自定义 UITableViewCell。

(1) 选中 AllShopViewController. m 文件，按下 command＋N，选择 iOS 下的 Objective-C class，单击 Next，如图 6.13 所示。

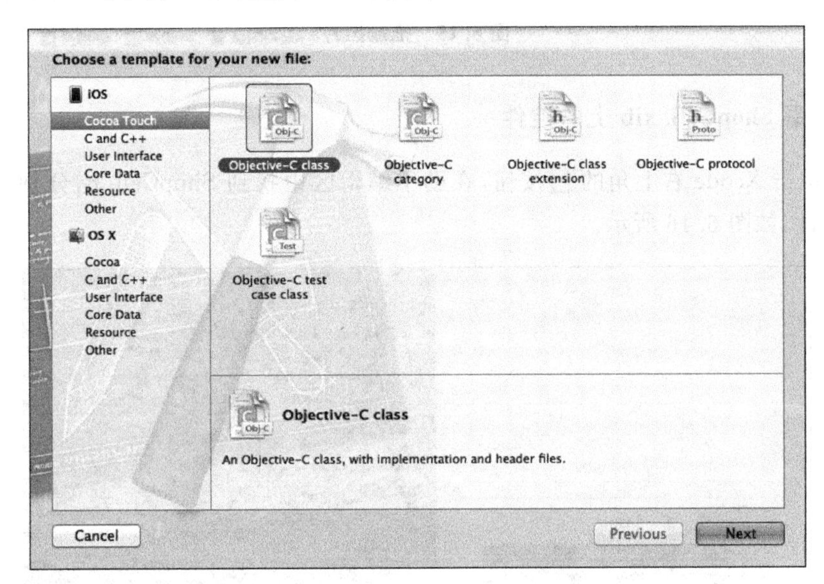

图 6.13　选择 Objective-C class

(2) 在 Class 中填写 ShopCell，在 Subclass of 中填写 UITableViewCell，此处勾选 Also create XIB file，单击 Next，如图 6.14 所示。

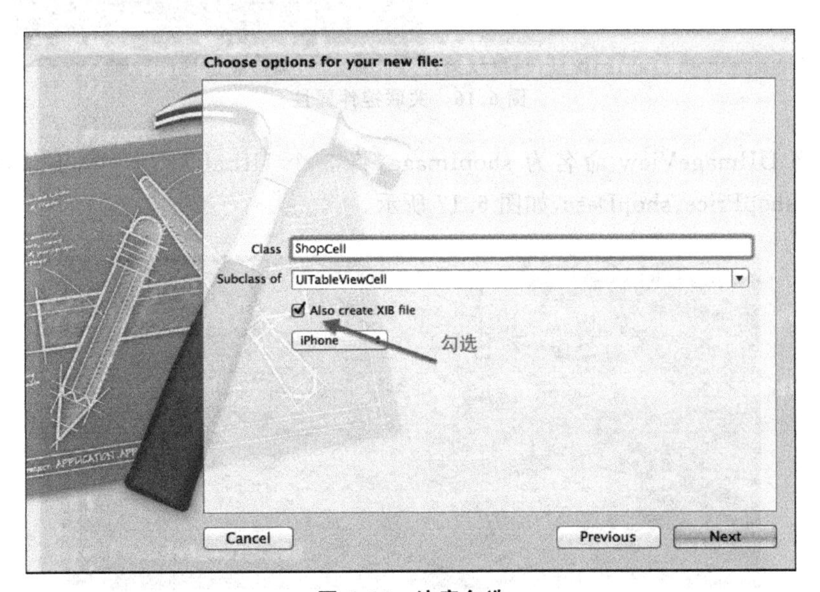

图 6.14　注意勾选

(3) 单击 ShopCell. xib。拖放 1 个 UIImageView 和 3 个 UILable 到 xib 上，如图 6.15 所示。

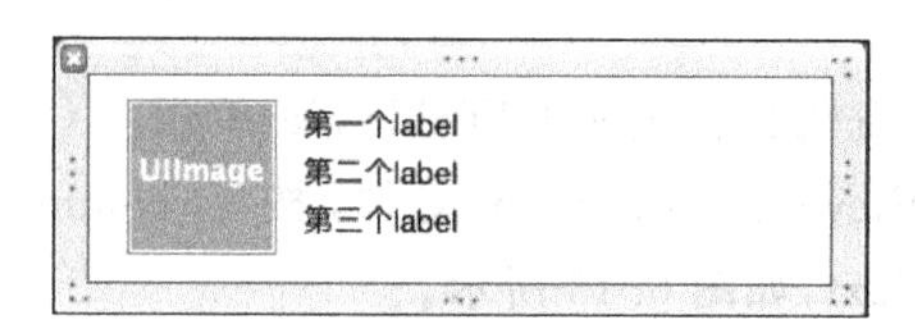

图 6.15 拖放控件

5. 关联 ShopCell. xib 上的控件

(1) 单击 Xcode 右上角的按钮，在助手编辑区中找到 ShopCell. h，分别关联 4 个控件的属性，如图 6.16 所示。

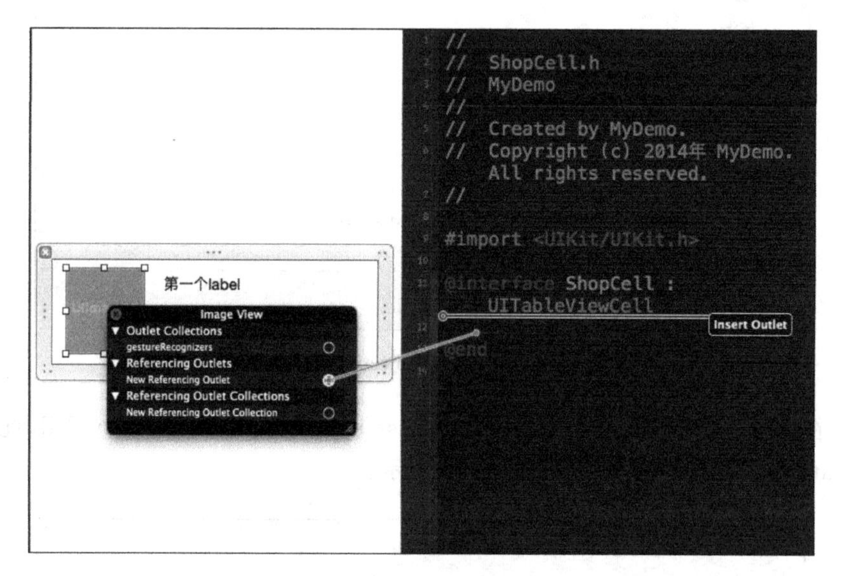

图 6.16 关联控件属性

(2) 将 UIImageView 命名为 shopImage，将 3 个 UILablel 从上往下依次命名为 shopTitle、shopPrice、shopDesc，如图 6.17 所示。

```
#import <UIKit/UIKit.h>

@interface ShopCell : UITableViewCell
@property (weak, nonatomic) IBOutlet UIImageView *shopImage;
@property (weak, nonatomic) IBOutlet UILabel *shopTitle;
@property (weak, nonatomic) IBOutlet UILabel *shopPrice;
@property (weak, nonatomic) IBOutlet UILabel *shopDesc;

@end
```

图 6.17 分别命名

(3) 将 tableView 的 cell 更改为 ShopCell。单击 AllShopViewController. m,更改代码如下：

```
#import "AllShopViewController.h"
#import "ShopCell.h"
@interface AllShopViewController ()<UITableViewDataSource,UITableViewDelegate>
@property (nonatomic,strong)UITableView * table;
@end

@implementation AllShopViewController

- (void)loadView{
    [super loadView];
    self.title=@"全部商户";
    _table=[[UITableView alloc]initWithFrame:CGRectMake(0, 0, App_SCREEN_
WIDTH, App_SCREEN_HEIGHT-64) style:UITableViewStylePlain];
    [self.view addSubview:_table];
    _table.rowHeight=86;
    _table.dataSource=self ;
    _table.delegate=self;
}
- (void)viewDidLoad
{
    [super viewDidLoad];
}
- (void)leftBarChick{
    [self dismissViewControllerAnimated:YES completion:nil];
}
- (NSInteger) tableView: (UITableView * ) tableView numberOfRowsInSection:
(NSInteger)section{
    return5;  //由于没有对接服务端,暂时返回 5 行
}
- (UITableViewCell *)tableView:(UITableView *)tableView cellForRowAtIndexPath:
(NSIndexPath *)indexPath{
    //指定 cellIdentifier 为自定义的 cell
    static NSString * CellIdentifier=@"ShopCell";
    //自定义 cell 类
    ShopCell * cell=[tableView dequeueReusableCellWithIdentifier:
CellIdentifier];
    if (cell==nil) {
        //通过 xib 的名称加载自定义的 cell
        cell=[[[NSBundle mainBundle] loadNibNamed:@"ShopCell" owner:self
options:nil] lastObject];
    }
    cell.shopImage.image=[UIImage imageNamed:@"default.png"];
```

```
    cell.shopTitle.text=@"名字";
    cell.shopDesc.text=@"描述";
    cell.shopPrice.text=@"人均消费";
    return cell;
}
-(void)tableView:(UITableView *)tableView didSelectRowAtIndexPath:(NSIndexPath
*)indexPath{
}

-(void)didReceiveMemoryWarning
{
    [super didReceiveMemoryWarning];
    //Dispose of any resources that can be recreated.
}

@end
```

代码解析：

注意 tableView：cellForRowAtIndexPath：方法中，用到了自定义 shopCell，自定义 shopCell 也遵循 UITableViewCell 的复用机制。tableView 首先查找是否有可重用的 cell，如果有则返回，没有则创建自定义 cell。创建自定义 cell 时，程序会寻找名为 shopCell 的 xib 文件并返回。

NSBundle 是 cocoa 提供的一个类，表示是一个目录（bundle），其中包含了程序会使用的资源。这些资源包含了图像、声音、编译好的代码、xib 文件。

我们的程序是一个 bundle。在 Finder 中，一个应用程序看上去和其他文件没有什么区别，但是实际上它是一个包含了 xib 文件、编译代码以及其他资源的目录。这个目录就叫做程序的 mainbundle。通过使用下面的方法得到程序的 mainbundle：

```
NSBundle *myBundle=[NSBundle mainBundle];
```

6. 设置全部商户页面为启动页面

（1）选中 AppDelegate.m，导入头文件并重写 application：didFinishLaunchingWithOptions：方法，代码如下：

```
#import "AppDelegate.h"
#import "MFSideMenuContainerViewController.h"
#import "AllShopViewController.h"
#import "LeftViewController.h"
@implementation AppDelegate
```

```
- (BOOL) application: (UIApplication * ) application didFinishLaunchingWithOptions:
(NSDictionary * )launchOptions
{
     self.window = [[UIWindow alloc] initWithFrame: [[UIScreen mainScreen]
bounds]];
    //创建侧滑实例
    MFSideMenuContainerViewController * mfsilderVC = [[MFSideMenuContainer
ViewController alloc]init];
    AllShopViewController * vc= [[AllShopViewController alloc]init];
    vc.title=@"全部商户";
    UINavigationController * mainNav = [[UINavigationController alloc]
initWithRoot ViewController:vc];
    LeftViewController * leftVC= [[LeftViewController alloc]initWithNibName:@"
LeftViewController" bundle:nil];
    //设置侧滑框架左侧滑页
    [mfsilderVC setLeftMenuViewController:leftVC];
    //设置侧滑框架中心页面
    [mfsilderVC setCenterViewController:mainNav];
    //设置左侧滑页显示宽度
    mfsilderVC.leftMenuWidth=270.0f;
    self.window.rootViewController=mfsilderVC;
    self.window.backgroundColor= [UIColor whiteColor];
    [self.window makeKeyAndVisible];
    return YES;
}
```

代码解析：

这里需要实例化一个 MFSideMenuContainerViewController 对象，然后设置它的左侧滑页面、中心侧滑页面以及左侧滑页面显示的宽度。最后将其设置为 self. window 的 rootViewController。

(2) 单击 LeftViewController. m，修改代码如下：

```
#import "LeftViewController.h"
#import "MFSideMenuContainerViewController.h"
#import "LoginViewController.h"
#import "AllShopViewController.h"
@interface LeftViewController ()

@end

@implementation LeftViewController

- (id)initWithNibName:(NSString * )nibNameOrNil bundle:(NSBundle * ) nibBundleOrNil
```

```
{
    self=[super initWithNibName:nibNameOrNil bundle:nibBundleOrNil];
    if (self) {
    }
    return self;
}
-(void)dealloc{
    [NOTIFICATION_CENTER removeObserver:self];
}
-(void)viewDidLoad
{
    [super viewDidLoad];
    UIButton *button=(UIButton *)[self.view viewWithTag:100];
    if([USER_DEFAULT boolForKey:@"isLogin"]){
        [button setTitle:@"注销" forState:UIControlStateNormal];
    }else{
        [button setTitle:@"登录" forState:UIControlStateNormal];
    }
    [NOTIFICATION_CENTER addObserver:self selector:@selector(loginStateChange)
name:@"loginStateChange" object:nil];
}
-(void)loginStateChange{
    UIButton *button=(UIButton *)[self.view viewWithTag:100];
    if([USER_DEFAULT boolForKey:@"isLogin"]){
        [button setTitle:@"注销" forState:UIControlStateNormal];
    }else{
        [button setTitle:@"登录" forState:UIControlStateNormal];
    }
}
-(void)didReceiveMemoryWarning
{
    [super didReceiveMemoryWarning];
    //Dispose of any resources that can be recreated.
}

-(IBAction)goAllShop:(id)sender {
    AllShopViewController *vc=[[AllShopViewController alloc]init];
    UINavigationController * navigationController = self. menuContai-
nerViewController.centerViewController;
    NSArray *controllers=[NSArray arrayWithObject:vc];
    navigationController.viewControllers=controllers;
    [self.menuContainerViewController setMenuState:MFSideMenuStateClosed];
}
```

```
- (IBAction)goLogin:(id)sender {
    if([USER_DEFAULT boolForKey:@"isLogin"]){
        [USER_DEFAULT setBool:NO forKey:@"isLogin"];
        [USER_DEFAULT synchronize];
        UIAlertView * alert = [[UIAlertView alloc] initWithTitle: @"提示"
message:@"注销成功" delegate:nil cancelButtonTitle:@"好的" otherButto-
nTitles:nil, nil];
        [alert show];
        [NOTIFICATION_ CENTER postNotificationName: @" loginStateChange "
object:nil];
    }else{
        LoginViewController * vc= [[LoginViewController alloc] initWithNib
Name:@"LoginViewController" bundle:nil];
        vc.title=@"登录";
        UINavigationController * nav= [[UINavigationController alloc] initWi-
thRootViewController:vc];
        [self presentViewController:nav animated:YES completion:nil];
    }
}

#pragma MFSideMenuContainerViewController method
- (MFSideMenuContainerViewController * )menuContainerViewController {
    return(MFSideMenuContainerViewController * )self.parentViewController;
}

@end
```

代码解析：

在 viewDidLoad 方法中，程序用到了 NSNotificationCenter(通知)。

程序开发的时候会遇到这种情况：打开 App 后，程序后台运行某个方法，例如下载文件，下载完成后可能需要调用某个方法来刷新界面，这时没法在下载的函数中回调。NSNotificationCenter(通知)是一个很好的选择。在调用通知时程序会在整个项目中寻找此通知的名称，找到后发出请求，因此通知的名称需要在整个项目中是唯一的。

通知使用起来非常简单。定义通知，代码如下：

```
[[NSNotificationCenter defaultCenter] addObserver: self selector: @selector
(someMethod) name: @"someMethod" object: nil];
```

实现通知中使用的方法，代码如下：

```
- (void)someMethod{
}
```

调用通知,代码如下:

```
[[NSNotificationCenter defaultCenter] postNotificationName:@" someMethod "
object:self];
```

清除通知,代码如下:

```
- (void)dealloc{
    [[NSNotificationCenter defaultCenter] removeObserver:self];
}
```

结合本项目,当用户登录成功,程序首先保存用户信息,然后发送一个登录状态改变的通知,将左侧滑栏登录按钮文字替换为"注销"。当用户单击注销时,程序首先删除登录状态信息,再发送一个登录状态改变的通知,将注销按钮文字替换为"登录"。

(3) 单击 LoginViewController. m,登录成功后发送通知,代码如下:

```
- (IBAction)loginChick:(id)sender {
    if(self.username.text.length<1||self.password.text.length<1){
        UIAlertView * alert= [[UIAlertView alloc]initWithTitle:@"提示" message:
@"请完整输入" delegate:nil cancelButtonTitle:@"好的" otherButtonTitles:nil,
nil];
        [alert show];
    }
    else{
        NSDictionary * dic=@{@"act":@"login",@"username":_username.text,
@"password":_password.text};
        [AFHelper downDataWithDictionary: dic andBaseURLStr: @" http:
//localhost:8080/meServer/" andPostPath:@"user.php?" success:^(NSDictio-
nary * dic){
            if([dic[@"flag"]intValue]==1){
                NSLog(@"登录成功");
                [USER_DEFAULT setBool:YES forKey:@"isLogin"];
                [USER_DEFAULT setObject:_username.text forKey:@"username"];
                [USER_DEFAULT synchronize];
                //发送通知
                [NOTIFICATION_CENTER postNotificationName: @" loginStateChange"
object:nil];
                [self dismissViewControllerAnimated:YES completion:nil];
            }else{
                UIAlertView * alert= [[UIAlertView alloc]initWithTitle:@"提示"
message:dic[@"msg"] delegate:nil cancelButtonTitle:@"好的" otherButtonTitles:
nil, nil];
```

```
                [alert show];
            }
        }];
    }
}
```

6.3.2 详情内容展示

单击全部商户中的行，程序会跳转到该商户的详情页，下面来添加商户详情页。

1. 添加商户详情页

（1）选中 Section 文件夹，按下 command+N，选择 iOS 下的 Objective-C class，如图 6.18 所示。

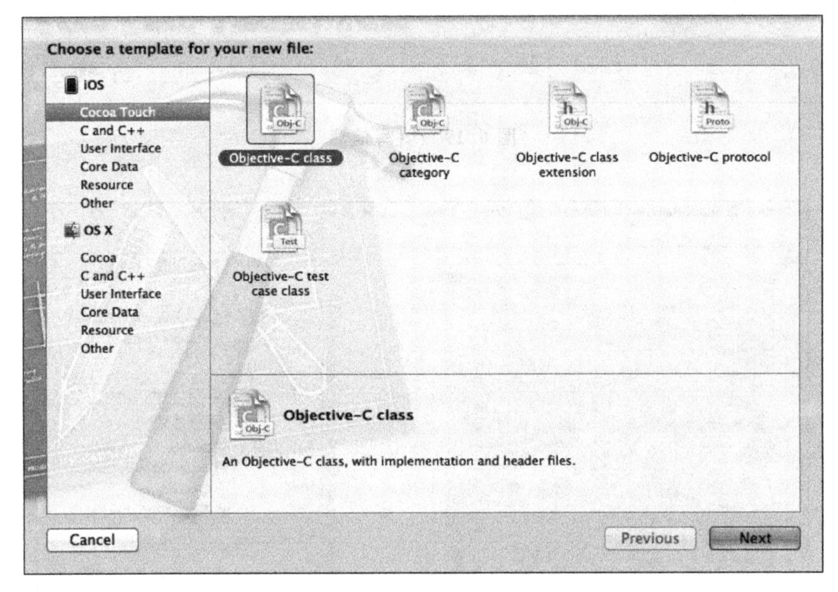

图 6.18 选择 Objective-C class

（2）在 Class 中填写 ShopDetailViewController，在 Subclass of 中填写 FatherViewController，勾选 Also create XIB file，如图 6.19 所示。

（3）选中 ShopDetailViewController.xib，拖放一个 View 到 xib 上，如图 6.20 所示。这里添加 View 是为了区分 xib 的背景颜色，没有其他作用。

（4）分别拖放 5 个 UILable、5 个 UIImageView、3 个 UIButton 到 View 上，并调整布局，如图 6.21 所示。

（5）设置 Label1 字体大小为 System 15.0，设置 Label2、Label3、Label4、Label5 字体大小为 System 13.0，颜色为 Dark Gray Color，如图 6.22 所示。

（6）设置 Button1 和 Button2 属性，如图 6.23 所示。

图 6.19　填写信息

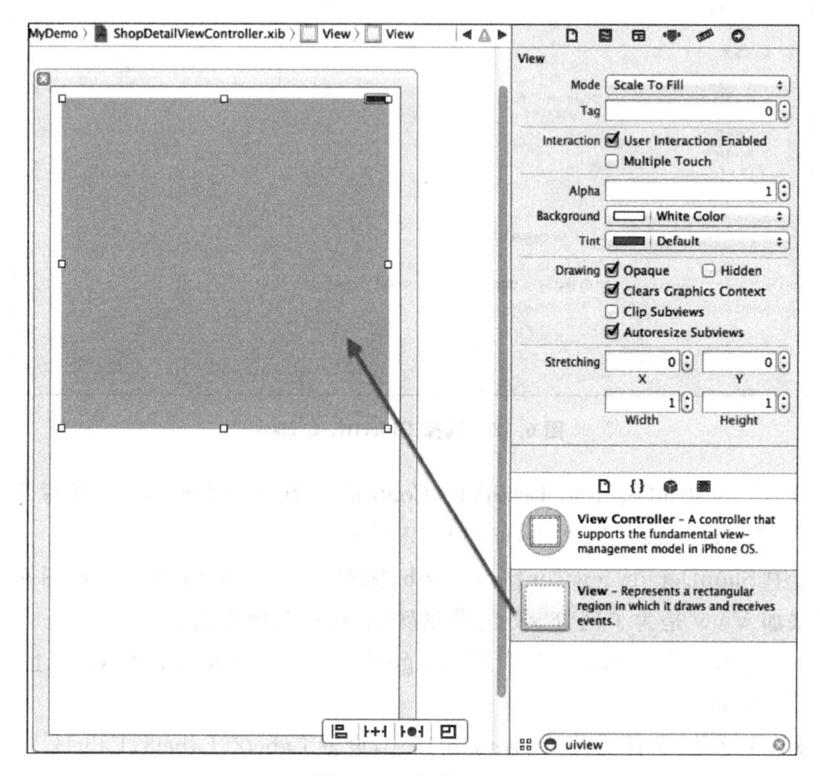

图 6.20　拖放 View

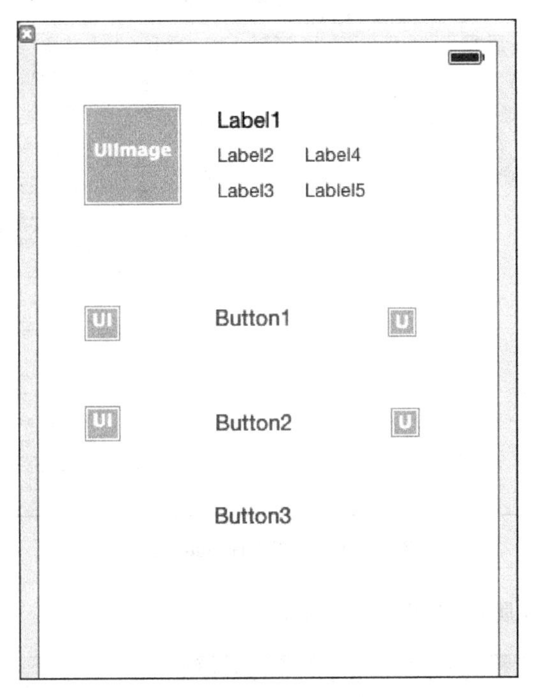

图 6.21 拖放控件

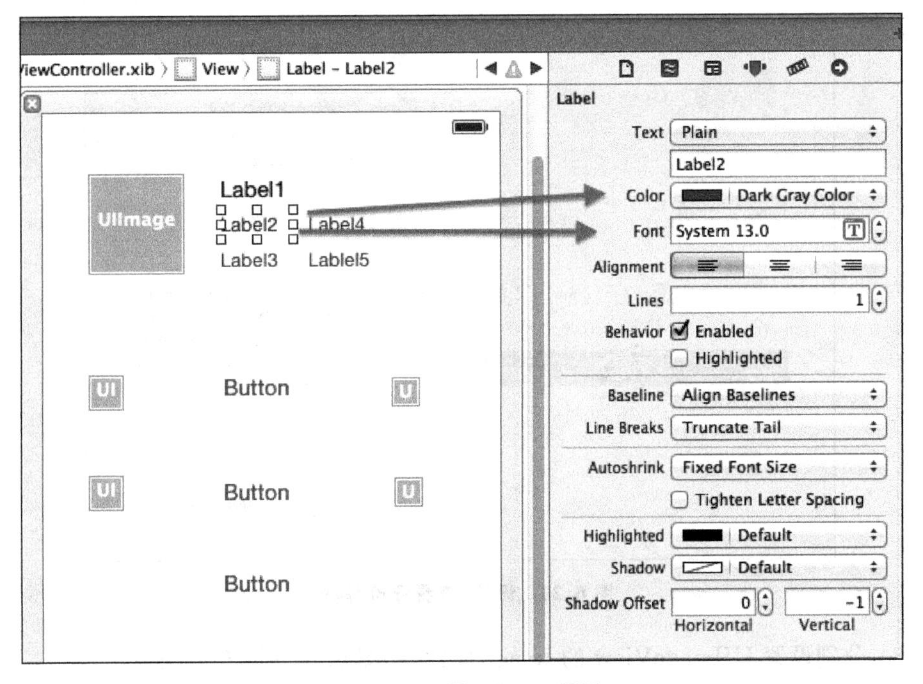

图 6.22 设置 UILabel 属性

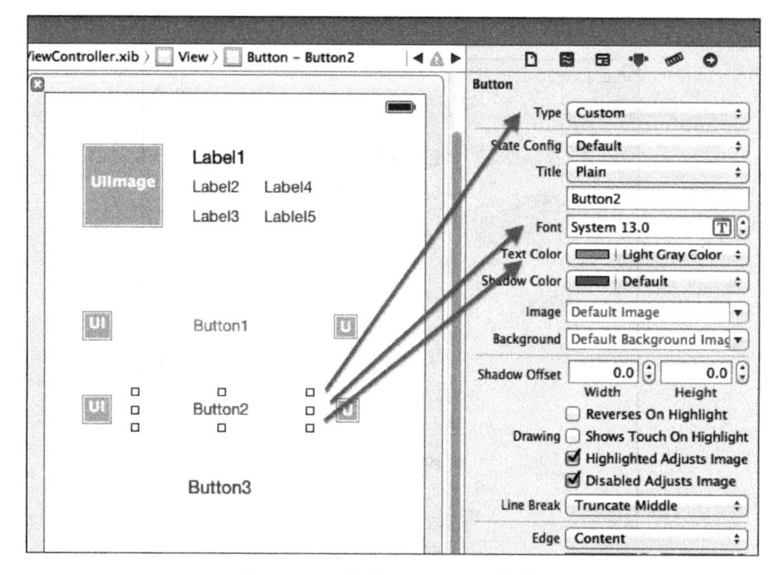

图 6.23　设置 UIBtuuon 属性

(7) 设置 Button3 属性,如图 6.24 所示。

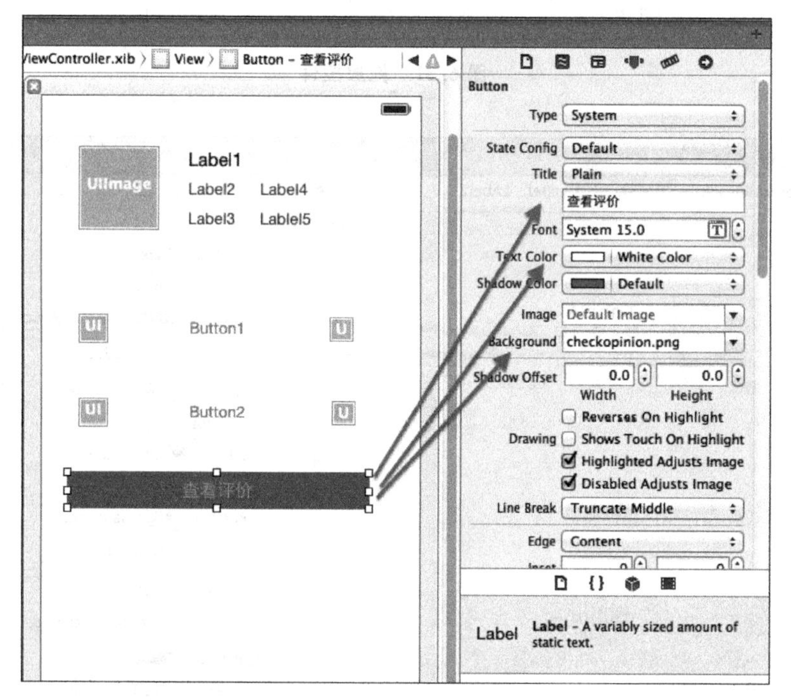

图 6.24　设置“查看评价”属性

(8) 分别设置 UIImageView 的 Image 属性,如图 6.25～图 6.27 所示。

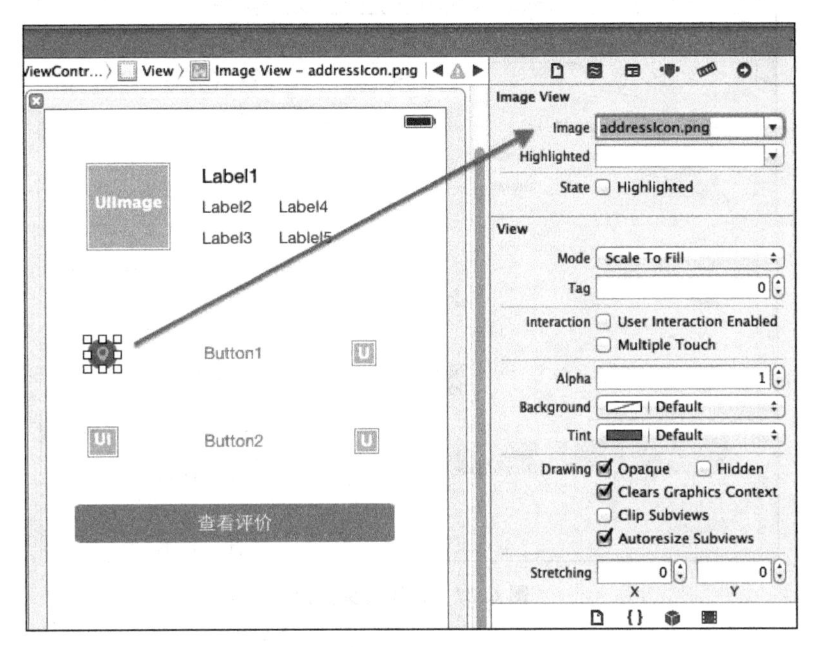

图 6.25 设置位置图片

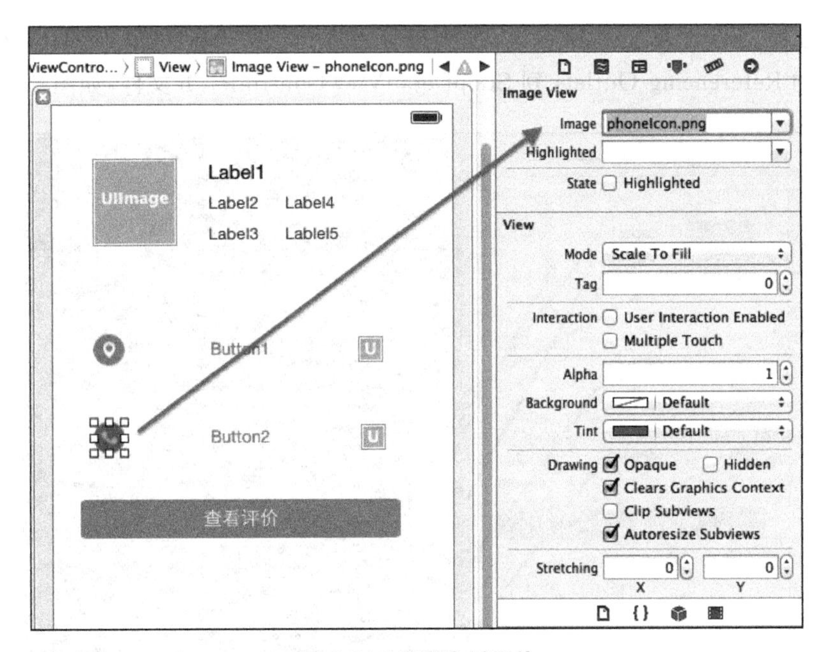

图 6.26 设置电话图片

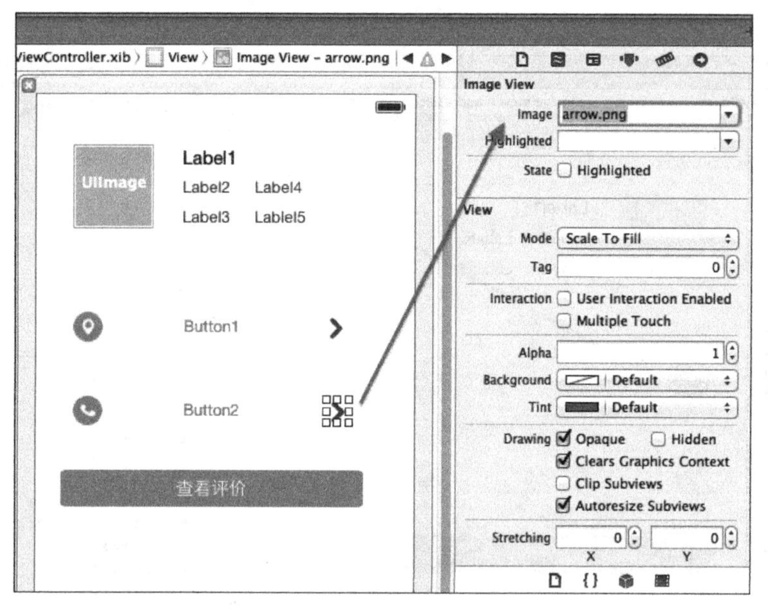

图 6.27　设置箭头图片

2. 关联 ShopDetailViewController.xib 控件

(1) 设置 Label2 和 Label3 的 Text 属性分别为"距离:"和"营业时间:",然后分别关联控件的 Referencing Outlets 到 ShopDetailViewController.h 文件,如图 6.28 所示。

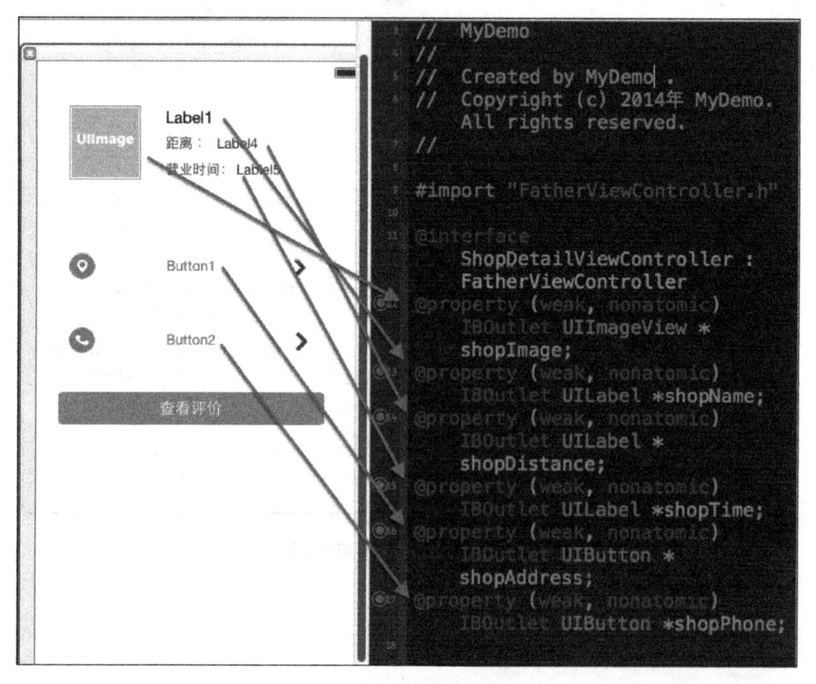

图 6.28　关联控件属性

(2) 分别关联 UIButton 控件的 Touch Up Inside 事件并命名,如图 6.29 所示。

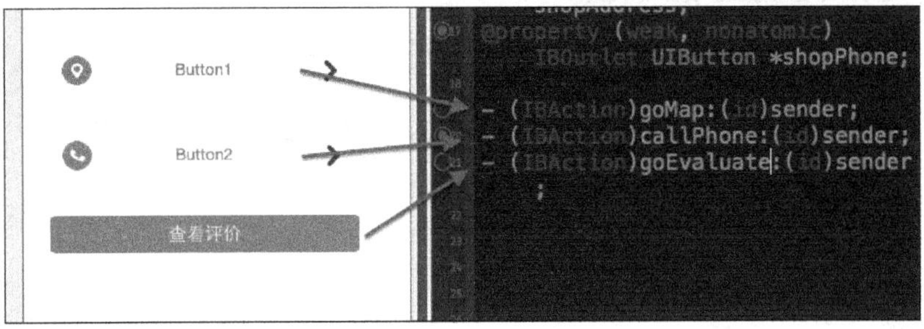

图 6.29 关联控件事件

6.3.3 客户端与服务端交互

在开始编写代码前,先介绍 iOS 程序设计中非常重要的"模型-视图-控制器(MVC)"范型。

MVC 把软件系统分为 3 部分: 模型(Model)、视图(View)、控制器(Controller)。在 iOS 中,程序中的每一个 Object(对象)都明显地仅属于这 3 部分中的某一个。下面举例说明。

- 模型

举个例子,上中学的时候,步步高电子词典中有个游戏叫"雷霆战机",也就是"打飞机"的游戏。模型就是: 小飞机的攻击力是多少? 小飞机上装的是什么武器,是炮弹、导弹,还是激光炮? 小飞机还有多少血? 等等。概括地说,就是程序将要实现的功能,或者是它所能干的事情。

- 控制器

控制器是程序内部的逻辑,大多情况下看不到它。它将模型和视图捆绑在一起,处理用户的输入。例如,按开炮键后,控制器就会通过内部的逻辑来处理要求,并在屏幕上做出相应的显示,我们可看到屏幕上的小飞机发出炮弹击中了敌机。这是控制器控制视图显示的例子,可以把控制器看成是连接模型和视图的桥梁。

- 视图

视图就是小飞机是什么样子的。它有一个还是两个翅膀,有几挺枪炮;飞机在屏幕上的位置如何等。总之,屏幕上看到的组件都可以归类为视图。

下面结合本项目,介绍如何采用 MVC 模式进行软件开发。

1. 添加 Model 类

(1) 右击项目左侧的 MyDemo 文件夹,单击 New Group,将新建文件夹命名为 Models。

(2) 选中 Models 文件夹,按下 command+N,选择 Cocoa Touch 下的 Objective-C class,如图 6.30 所示。

(3) 在 Class 中输入 ShopItem,Subclass of 输入 NSObject,如图 6.31 所示。

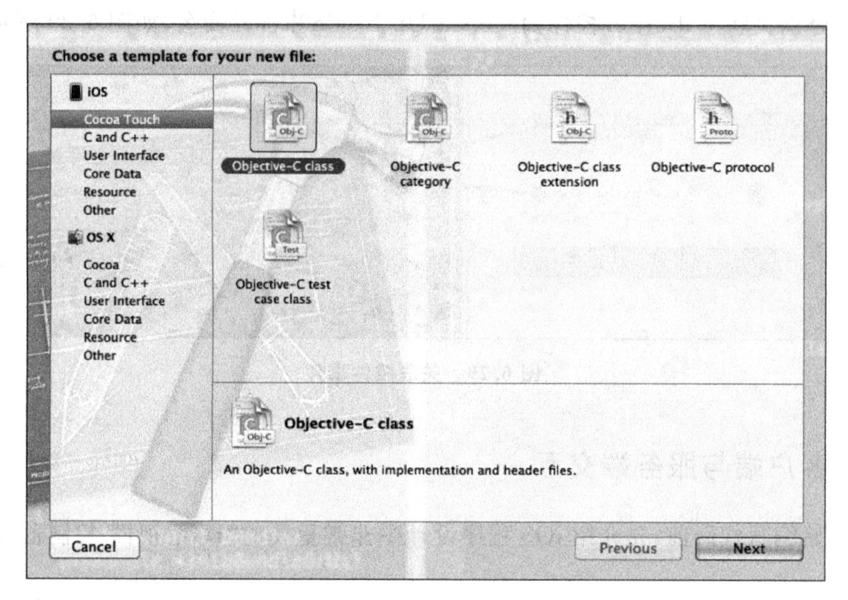

图 6.30 选择 Objective-C class

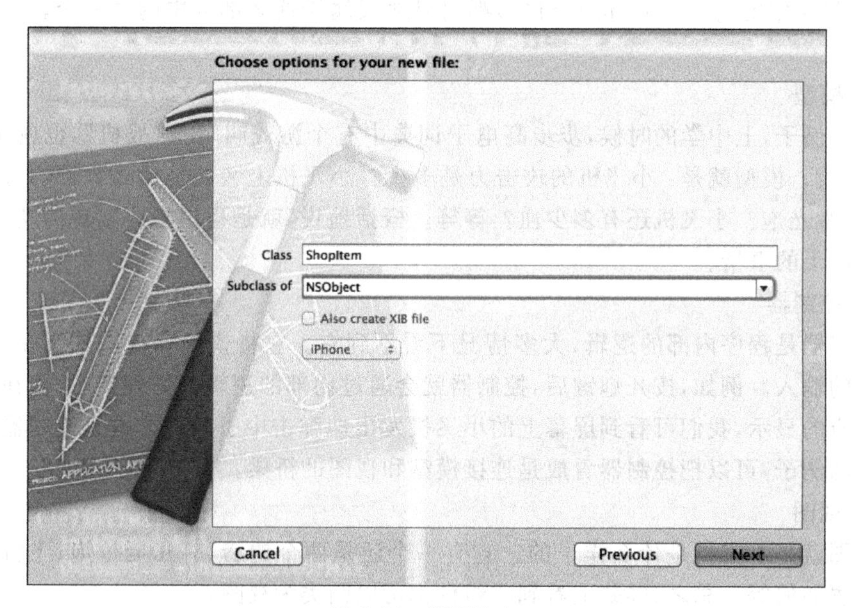

图 6.31 填写信息

(4) 选择保存路径后，单击 ShopItem.h。修改代码如下：

```
#import<Foundation/Foundation.h>

@interface ShopItem: NSObject

@property  (nonatomic,copy) NSString * shopId;
```

```
@property  (nonatomic,copy) NSString * shopAddress;
@property  (nonatomic,copy) NSString * shopDesc;
@property  (nonatomic,copy) NSString * shopImage;
@property  (nonatomic,copy) NSString * shopLatitude;
@property  (nonatomic,copy) NSString * shopLongitude;
@property  (nonatomic,copy) NSString * shopName;
@property  (nonatomic,copy) NSString * shopPhone;
@property  (nonatomic,copy) NSString * shopSpend;
@property  (nonatomic,copy) NSString * shopTime;
@property  (nonatomic,copy) NSString * shopType;
@property  (nonatomic,copy) NSString * shopDistance;

@end
```

代码解析：

添加属性用来接收接口返回对应字段的值。此处不需要在.m文件中实现set和get方法，Xcode会自动加上。

这里的字符串的属性用到的是copy，为什么要用copy呢？因为如果用retain，则不能确定调用者传的是NSString还是NSMutableString。如果传的是NSMutableString，那么这个NSString属性内容就有可能受调用者的影响，而NSString通常不希望被调用者改变。

2. 修改AllShopViewController代码

(1) 单击AllShopViewController.h，添加数组，用来装载数据模型类，代码如下：

```
#import "FatherViewController.h"
@interface AllShopViewController: FatherViewController
@property (nonatomic,strong)NSMutableArray * dataArray;
@end
```

代码解析：

上面代码中，程序创建一个可变数组dataArray，用来装载模型对象。在介绍可变数组之前，先介绍数组(NSArray)。

NSArray是不可变的数组，只能储存Objective-C对象。如果想保存一些原始的C数据(例如int、float、double等)，则需要将这些原始的C数据封装成NSNumber类型(NSNumber对象是cocoa对象，可以被保存在集合类中)，然后再添加到数组中。

对数组进行的操作可以分为增(插入)、删、改、查、取、遍历、排序、比较等。对于NSArray来说，因为其本身是不可变的，所以对NSArray的操作没有增和删。如果要进行增和删的操作，那么就要用到可变数组(NSMutableArray)。NSMutableArray继承于NSArray，除了拥有NSArray的方法外，它还允许直接添加、删除、交换元素等操作。因此在不确定输入元素个数的时候，应该用NSMutableArray。

(2) 单击 AllShopViewController.m,修改代码如下：

```
#import "AllShopViewController.h"
#import "ShopCell.h"
#import "ShopItem.h"
#import "AFHelper.h"
#import "UIImageView+AFNetworking.h"
@interface AllShopViewController ()<UITableViewDataSource,UITableViewDelegate>
@property (nonatomic,strong)UITableView * table;
@end

@implementation AllShopViewController

- (void)loadView{
    [super loadView];
    _table= [[UITableView alloc]initWithFrame:CGRectMake(0, 0, App_SCREEN_
WIDTH, App_SCREEN_HEIGHT-64) style:UITableViewStylePlain];
    [self.view addSubview:_table];
    _table.rowHeight=86;
    _table.dataSource=self ;
    _table.delegate=self;
}
- (NSMutableArray * ) dataArray {
    if(!_dataArray){
        _dataArray= [[NSMutableArray alloc]init];
    }
    return _dataArray;}
- (void)viewDidLoad
{
    [super viewDidLoad];
    [self createLeftMenuBtn];
    [self getData];
}
- (void)getData{
    NSDictionary * dic=@{@"act":@"getShops",@"pageSize":@"20",@"page":@
"1",@"latitude":@"30.562710",@"longitude":@"104.075151"};
    [AFHelper downDataWithDictionary:dic andBaseURLStr:@"http://localhost:
8080/meServer/" andPostPath:@"shop.php?" success:^(NSDictionary * dic){
        if([dic[@"flag"]intValue]==1){
            for(NSDictionary * tmpDic in dic[@"vendorList"]){
                ShopItem * shopItem= [[ShopItem alloc]init];
                shopItem.shopId=tmpDic[@"shopId"];
                shopItem.shopAddress=tmpDic[@"shopAddress"];
```

```
                shopItem.shopDesc=tmpDic[@"shopDesc"];
                shopItem.shopImage=tmpDic[@"shopImage"];
                shopItem.shopLatitude=tmpDic[@"shopLatitude"];
                shopItem.shopLongitude=tmpDic[@"shopLongitude"];
                shopItem.shopName=tmpDic[@"shopName"];
                shopItem.shopPhone=tmpDic[@"shopPhone"];
                shopItem.shopSpend=tmpDic[@"shopSpend"];
                shopItem.shopTime=tmpDic[@"shopTime"];
                shopItem.shopType=tmpDic[@"shopType"];
                shopItem.shopDistance=tmpDic[@"distance"];
                [self.dataArray addObject:shopItem];
            }
            [_table reloadData];
        }else{
            UIAlertView * alert = [[UIAlertView alloc] initWithTitle:@"提示"
message:dic[@"msg"] delegate:nil cancelButtonTitle:@"好的" otherButtonTitles:
nil, nil];
            [alert show];
        }
    }];
}
- (NSInteger) tableView: (UITableView * ) tableView numberOfRowsInSection:
(NSInteger)section{
    return self.dataArray.count;
}
- (UITableViewCell * ) tableView: (UITableView * ) tableView cellForRowAt
IndexPath:(NSIndexPath *)indexPath{
    //指定 cellIdentifier 为自定义的 cell
    static NSString * CellIdentifier=@"ShopCell";
    //自定义 cell 类
    ShopCell * cell=[tableView dequeueReusableCellWithIdentifier:CellIdentifier];
    if (cell==nil) {
        //通过 xib 的名称加载自定义的 cell
         cell= [[[NSBundle mainBundle] loadNibNamed:@"ShopCell" owner:self
options:nil] lastObject];
    }
    ShopItem * shopItem=_dataArray[indexPath.row ];
    [cell.shopImage setImageWithURL: [ NSURL URLWithString: shopItem.
shopImage] placeholderImage:nil];
    cell.shopTitle.text=shopItem.shopName;
    cell.shopPrice.text=[NSString stringWithFormat:@"人均: %@元",shopItem.
shopSpend];
```

```
    cell.shopDesc.text=shopItem.shopDesc;
    return cell;
}
- (void) tableView: (UITableView * ) tableView didSelectRowAtIndexPath:
(NSIndexPath * )indexPath{
}
- (void)didReceiveMemoryWarning
{
    [super didReceiveMemoryWarning];
}
@end
```

代码解析：

```
NSDictionary * dic=@{@"act":@"getShops",@"pageSize":@"20",@"page":@"1",@
"latitude":@"30.562710",@"longitude":@"104.075151"};
```

由于显示商户信息接口要返回商户和用户的距离，为方便讲解，这里固定传两个经纬度坐标。第7章将讲解定位相关知识，比如如何获取用户当前坐标。

```
for(NSDictionary * tmpDic in dic[@"vendorList"]){
    shopItem * shopItem=[[ShopItem alloc]init];
    shopItem.shopId=tmpDic[@"shopId"];
    shopItem.shopAddress=tmpDic[@"shopAddress"];
    shopItem.shopDesc=tmpDic[@"shopDesc"];
    shopItem.shopImage=tmpDic[@"shopImage"];
    shopItem.shopLatitude=tmpDic[@"shopLatitude"];
    shopItem.shopLongitude=tmpDic[@"shopLongitude"];
    shopItem.shopName=tmpDic[@"shopName"];
    shopItem.shopPhone=tmpDic[@"shopPhone"];
    shopItem.shopSpend=tmpDic[@"shopSpend"];
    shopItem.shopTime=tmpDic[@"shopTime"];
    shopItem.shopType=tmpDic[@"shopType"];
    shopItem.shopDistance=tmpDic[@"distance"];
    [self.dataArray addObject:shopItem];
}
```

服务端返回商户信息后，这里将我们所关心的字段解析出来，填充到Shopitem模型类对应的属性中，最后将模型类添加到dataArray中。

```
- (UITableViewCell * ) tableView: (UITableView * ) tableView cellForRowAt
IndexPath:(NSIndexPath * )indexPath{
    //指定 cellIdentifier 为自定义的 cell
```

```
    static NSString * CellIdentifier=@"ShopCell";
    //自定义 cell 类
    ShopCell * cell=[tableView dequeueReusableCellWithIdentifier:CellIdentifier];
    if (cell==nil) {
        //通过 xib 的名称加载自定义的 cell
        cell=[[[NSBundle mainBundle] loadNibNamed:@"ShopCell" owner:self
options:nil] lastObject];
    }
    ShopItem * shopItem=_dataArray[indexPath.row];
    [cell.shopImage setImageWithURL: [NSURL URLWithString: shopItem.
shopImage] placeholderImage:nil];
    cell.shopTitle.text=shopItem.shopName;
    cell.shopPrice.text=[NSString stringWithFormat:@"人均: %@元",shopItem.
shopSpend];
    cell.shopDesc.text=shopItem.shopDesc;
    return cell;
}
```

这里用到了 xib 自定义的 ShopCell。首先从 dataArray 中获取当前行对应的数据模型类,然后将数据模型类中的商户用户名、商户人均消费、商户描述传递给 cell 的控件,并显示出来。

3. 为商户详情页增加入口

(1) 单击 ShopDetailViewController.h,添加 ShopItem 属性,代码如下:

```
#import "FatherViewController.h"
#import "ShopItem.h"
@interface ShopDetailViewController: FatherViewController
@property (weak, nonatomic) IBOutlet UIImageView * shopImage;
@property (weak, nonatomic) IBOutlet UILabel * shopName;
@property (weak, nonatomic) IBOutlet UILabel * shopDistance;
@property (weak, nonatomic) IBOutlet UILabel * shopTime;
@property (weak, nonatomic) IBOutlet UILabel * shopAddress;
@property (weak, nonatomic) IBOutlet UILabel * shopPhone;
@property (strong,nonatomic)ShopItem * shopItem;
-(IBAction)goMapVC:(id)sender;
-(IBAction)callPhone:(id)sender;
-(IBAction)goEvaluateVC:(id)sender;
@end
```

代码解析:

ShopItem 是之前建立的模型类,用来映射商户的信息。当用户单击全部商户页中的行后，就会将该行的商户信息传递到商户详情页，因此，这里需要新建一个 ShopItem 来接收商户信息。

(2) 单击 AllShopViewController.m，导入 ShopDetailViewController 头文件，代码如下：

```
#import "AllShopViewController.h"
#import "ShopDetailViewController.h"
```

(3) 重写 tableView: didSelectRowAtIndexPath:方法，代码如下：

```
-(void)tableView:(UITableView *)tableView didSelectRowAtIndexPath:(NSIndexPath *)indexPath{
    ShopDetailViewController *vc = [[ShopDetailViewController alloc] initWithNibName:@"ShopDetailViewController" bundle:nil];
    vc.shopItem=_dataArray[indexPath.row];
    UINavigationController *nav=[[UINavigationController alloc] initWithRootViewController:vc];
    [self presentViewController:nav animated:YES completion:nil];
}
```

代码解析：

单击 UITableView 中的行时，会触发 tableView: didSelectRowAtIndexPath:方法，跳转到详情页，并将当前行的商户信息模型类传递给详情页的模型类。

(4) 单击 ShopDetailViewController.m，修改代码如下：

```
#import "ShopDetailViewController.h"
#import "UIImageView+AFNetworking.h"
@interface ShopDetailViewController ()

@end

@implementation ShopDetailViewController

-(id)initWithNibName:(NSString *)nibNameOrNil bundle:(NSBundle *)nibBundleOrNil
{
    self=[super initWithNibName:nibNameOrNil bundle:nibBundleOrNil];
    if (self) {
        //Custom initialization
    }
    return self;
}
```

```
- (void)viewDidLoad
{
    [super viewDidLoad];
    self.title=@"商户详情";
    [self createLeftBtnWithFrame:CGRectMake(0.0f, 0.0f, 24.0f, 24.0f) Image:
[UIImage imageNamed:@"backButtonNormal"]];
    [_shopImage setImageWithURL: [NSURL URLWithString: _shopItem.
shopImage]];
    _shopName.text=_shopItem.shopName;
    _shopDistance.text=[NSString stringWithFormat:@"距离: %@米",_shopItem.
shopDistance];
    _shopTime.text=_shopItem.shopTime;
    _shopAddress.text=_shopItem.shopAddress;
    _shopPhone.text=_shopItem.shopPhone;
}
- (void)leftBarChick{
    [self dismissViewControllerAnimated:YES completion:nil];
}
- (void)didReceiveMemoryWarning
{
    [super didReceiveMemoryWarning];
    //Dispose of any resources that can be recreated.
}
- (IBAction)goMapVC:(id)sender {
}
- (IBAction)callPhone:(id)sender {
    [[UIApplication sharedApplication] openURL: [NSURL URLWithString:
[NSString stringWithFormat:@"tel://%@",_shopItem.shopPhone]]];
}
- (IBAction)goEvaluateVC:(id)sender {
}
@end
```

代码解析：

setImageWithURL:方法是AFNetworking框架中加载网络图片的方法。只需要传一个图片地址url,就能在UIImageView上显示图片,非常方便。

```
- (IBAction)callPhone:(id)sender {
    [[UIApplication sharedApplication] openURL:[NSURL URLWithString:[NSString
stringWithFormat:@"tel://%@",_shopItem.shopPhone]]];
}
```

这个方法调用iOS系统拨打电话的功能。

类似的还有发短信的功能，代码如下：

```
[[UIApplication sharedApplication] openURL:[NSURL URLWithString:@"sms://
10086"]];
```

发邮件的功能，代码如下：

```
[[UIApplication sharedApplication] openURL: [NSURL URLWithString: @
"mailto://admin@apple.com"]];
```

调用自带的 safari 浏览器，代码如下：

```
[[UIApplication sharedApplication] openURL:[NSURL URLWithString:@"http://
www.apple.com"]];
```

4. 手动添加商户信息到数据库

由于第 2 章搭建的本地服务端数据库商户表中没有数据，为方便讲解，这里手动添加商户信息。有两种方式添加数据。

第一种方式：依次复制下列链接到浏览器并回车。

```
    http://localhost:8080/meServer/shop.php?act=postShop&shopAddress=成都高新
区天府软件园 A 区 12 号 &shopDesc=全店清仓,买一送三,多买多送,速来抢购 &shopImage=
http://www.hqls.com/images/logo.jpg&shopLatitude=30.552728&shopLongitude=
104.075181&shopName=红旗连锁 &shopPhone=4001235212&shopSpend=20&shopTime=9:000
-23:00&shopType=2
    http://localhost:8080/meServer/shop.php?act=postShop&shopAddress=成都
高新区软件园 D 区 1210&shopDesc=豪客来优惠券(全国版):凭此券消费西式套餐,省 6 元
&shopImage = http://info.moonbasa.com/files/2011/5/6/11 - 873080413086.
jpg&shopLatitude=30.553661&shopLongitude=104.074175&shopName=豪客来牛排
&shopPhone=4005654872&shopSpend=50&shopTime=9:00-23:00&shopType=1
    http://localhost:8080/meServer/shop.php?act=postShop&shopAddress=成都高新
区天府软件园 D 区 1232&shopDesc=浩沙作为一个时尚健康生活方式的传播者,将致力于创建
一个国际一流的引领时尚健康生活的健身产业集团 &shopImage=http://www.hosafitness.
com/images/logo.jpg&shopLatitude=30.552752&shopLongitude=104.075121&shopName
=浩沙健身 &shopPhone=4004562121&shopSpend=50&shopTime=9:00-20:00&shopType=2
    ahttp://localhost:8080/meServer/shop.php?act=postShop&shopAddress=成
都高新区天府软件园 C 区 1232&shopDesc=一串串被竹签穿起的美味,你能经受住诱惑么
&shopImage=http://www.cy8.com.cn/content/uplode/218/logo.jpg&shopLatitude=
30.552752 &shopLongitude = 104.075192&shopName = 玉林串串 &shopPhone =
4004562124&shopSpend=30&shopTime=9:00-20:00&shopType=1
```

第二种方式：打开 MySQL Workbench，执行下列 SQL 语句。

```
INSERT INTO 'me_server'.'me_shop' ('shop_address', 'shop_name', 'shop_time',
'shop_phone', 'shop_image', 'shop_longitude', 'shop_latitude', 'shop_type',
'shop_desc', 'shop_spend') VALUES ('成都高新区天府软件园 A 区 12 号', '红旗连锁
', '9:000-23:00', '4001235212', 'http://www.hqls.com/images/logo.jpg', '104.
075181', '30.552728', '2', '全店清仓,买一送三,多买多送,速来抢购了', '20');

INSERT INTO 'me_server'.'me_shop' ('shop_address', 'shop_name', 'shop_time',
'shop_phone', 'shop_image', 'shop_longitude', 'shop_latitude', 'shop_type',
'shop_desc', 'shop_spend') VALUES ('成都高新区软件园 D 区 1210', '豪客来牛排',
'9:000-23:00', '4005654872', 'http://info.moonbasa.com/files/2011/5/6/11-
873080413086.jpg', '104.074175', '30.553661', '1', '豪客来优惠券(全国版):凭此券
消费西式套餐,省 6 元', '50');

INSERT INTO 'me_server'.'me_shop' ('shop_address', 'shop_name', 'shop_time',
'shop_phone', 'shop_image', 'shop_longitude', 'shop_latitude', 'shop_type',
'shop_desc', 'shop_spend') VALUES ('成都高新区天府软件园 D 区 1232', '浩沙健身',
'9:00-20:00', '4004562121', 'http://www.hosafitness.com/images/logo.jpg',
'104.075121', '30.552752', '2', '浩沙作为一个时尚健康生活方式的传播者,将致力于
创建一个国际一流的引领时尚健康生活的健身产业集团', '50');

INSERT INTO 'me_server'.'me_shop' ('shop_address', 'shop_name', 'shop_time',
'shop_phone', 'shop_image', 'shop_longitude', 'shop_latitude', 'shop_type',
'shop_desc', 'shop_spend') VALUES ('成都高新区天府软件园 C 区 1232', '玉林串串',
'9:00-20:00', '4004562124', 'http://www.cy8.com.cn/content/uplode/218/logo.
jpg', '104.075192', '30.552752', '1', '一串串被竹签穿起的美味,你能经受住诱惑么',
'30');
```

添加完成后单击 Xcode 左上角的▶按钮,运行程序。

6.4 图片的处理与效果实现

本节将为 UIImageView 增加单击事件,并利用 UIScrollView 实现一个简易的图片浏览器,支持放大、缩小图片,如图 6.32 所示。

6.4.1 图片添加手势

iOS 系统在 3.2 版以后,为方便开发者使用一些常用的手势,提供了手势识别 UIGestureRecognizer 类。UIGestureRecognizer 类是抽象类,下面的子类是具体的手势:

UITapGestureRecognizer——Tap(点一下);

UIPinchGestureRecognizer——Pinch(二指往内或往外拨动,平时经常用到的缩放);

UIRotationGestureRecognizer——Rotation(旋转);

UISwipeGestureRecognizer——Swipe(滑动,快速移动);

图 6.32 图片浏览器

UIPanGestureRecognizer——Pan(拖移,慢速移动);

UILongPressGestureRecognizer——LongPress(长按)。

使用手势很简单,分为两步:

(1) 创建手势实例。创建手势时,指定一个回调方法。当手势开始、改变或结束时,回调方法被调用。

(2) 将手势添加到需要识别的 View 中。每个手势只对应一个 View,当屏幕触摸在 View 的边界内时,如果手势和预定的一样,就会触发回调方法。

结合本项目实例,下面为 UIImageView 添加一个单击手势。

单击 ShopDetailViewController.m,在 viewDidLoad 中添加如下代码:

```
-(void)viewDidLoad{
    ……//其他代码
    _shopImage.userInteractionEnabled=YES;
    UITapGestureRecognizer * gesture = [[UITapGestureRecognizer alloc]
initWithTarget:self action:@selector(imageChick)];
    [self.shopImage addGestureRecognizer:gesture];
}
-(void)imageChick{
}
```

代码解析:

userInteractionEnabled 是 UIView 的属性。该属性值为布尔类型,如属性本身的名称所示,用来决定 UIView 是否接受并响应用户的交互。由于 UIImageView 继承于

UIView,因此 UIImageView 也拥有 userInteractionEnabled 属性。

UIView 的 userInteractionEnabled 属性默认为 YES,而 UIImageView 对该属性进行了覆盖,默认为 NO。因此,如果要让 UIImageView 响应手势,就需要设置其 userInteractionEnabled 为 YES。

6.4.2 分页与翻页

要实现图片浏览的功能,需要用到 UIScrollView+UIPageController 的控件组合。下面介绍如何使用这个控件组合。

添加图片浏览页面

(1)单击 Sections 文件夹,按下 command+N,选择 Cocoa Touch 下的 Objective-C class,如图 6.33 所示。

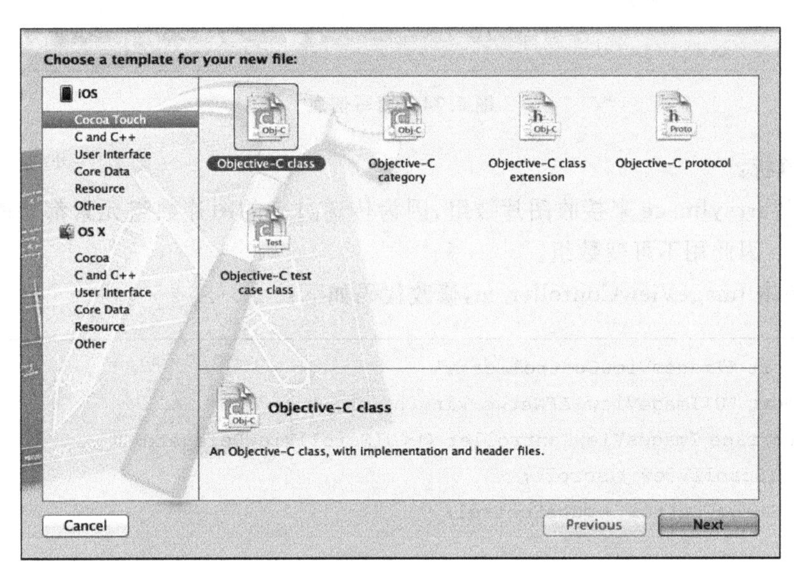

图 6.33 选择 Objective-C class

(2) 在 Class 中填写 ImagesViewController,在 Subclass of 中填写 UIViewController,不要勾选 Also create XIB file,如图 6.34 所示。

(3) 选择保存路径后,单击 ImageViewConroller. h,修改代码如下:

```
#import<UIKit/UIKit.h>

@interface ImagesViewController: UIViewController

@property(nonatomic,strong) NSArray * arrayImage;

@end
```

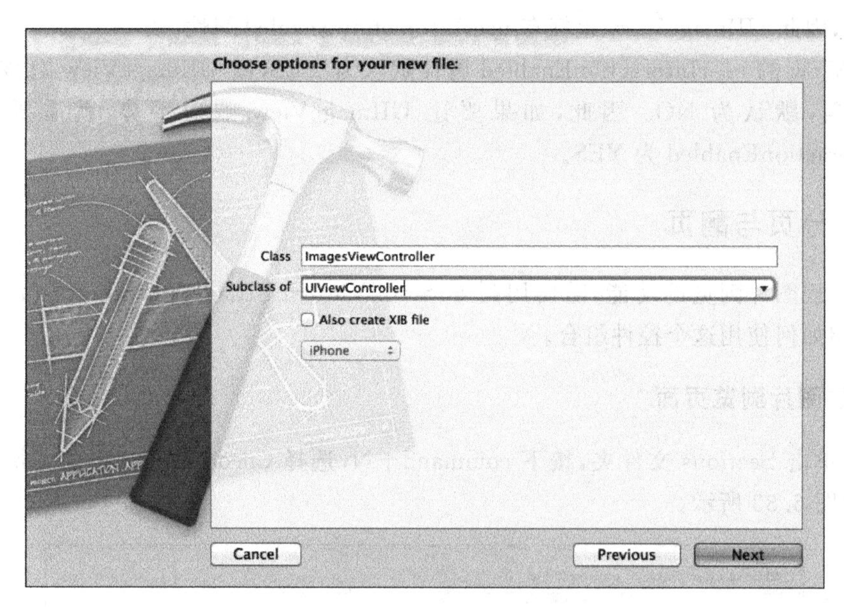

图 6.34 填写信息

代码解析：

这里用 arrayImage 来接收图片数组，因为传递过来的图片数组元素都是固定的，不会发生变化，因此用不可变数组。

(4) 单击 ImageViewConroller.m，修改代码如下：

```
#import "ImagesViewController.h"
#import "UIImageView+AFNetworking.h"
@interface ImagesViewController ()<UIScrollViewDelegate>{
    UIScrollView *scroll;
    UIPageControl *pageControl;
}
@end
@implementation ImagesViewController
-(void)loadView{
    [super loadView];
    self.view.backgroundColor=[UIColor blackColor];
    //创建一个 scrollView
    scroll=[[UIScrollView alloc]initWithFrame:CGRectMake(0.0f, 20.0f, 320.
0f, App_SCREEN_CONTENT_HEIGHT)];
     scroll.contentSize=CGSizeMake(320.0f*self.arrayImage.count, App_
SCREEN_CONTENT_HEIGHT);
    scroll.delegate=self;
    scroll.pagingEnabled=YES;
```

```
    scroll.bounces=NO;
    scroll.showsHorizontalScrollIndicator=NO;
    [self.view addSubview:scroll];
    //指示器
    pageControl= [[UIPageControl alloc] initWithFrame:CGRectMake(140.0f, App_
SCREEN_HEIGHT-20.0f, 40.0f, 15.0f)];
    [self.view addSubview:pageControl];
    pageControl.enabled=NO;
    pageControl.numberOfPages=self.arrayImage.count;
    for(int i=0; i<[self.arrayImage count];i++)
    {
        UIButton *button=[UIButton buttonWithType:UIButtonTypeCustom];
        button.frame=CGRectMake(0.0f, 0.0f, 320.0f, App_SCREEN_HEIGHT);
        [button addTarget:self action:@selector(chickImage) forControl-
Events:UIControlEventTouchUpInside];
        UIImageView *image= [[UIImageView alloc] initWithFrame:CGRectMake
(0.0f, 0.0f, 320.0f, App_SCREEN_HEIGHT)];
        [button addSubview:image];
        [image setContentMode:UIViewContentModeScaleAspectFit];
        NSString *imageStr = [NSString stringWithFormat:@"%@", [self.
arrayImage objectAtIndex:i]];
        [image setImageWithURL:[NSURL URLWithString:imageStr]];
        UIScrollView *pinchScrollView=[[UIScrollView alloc]initWithFrame:
CGRectMake(320.0f*i, 0.0f, 320.0f, App_SCREEN_HEIGHT)];
        pinchScrollView.delegate=self;
        pinchScrollView.maximumZoomScale=2.5;
        pinchScrollView.minimumZoomScale=0.5;
        [pinchScrollView addSubview:button];
        [scroll addSubview:pinchScrollView];
    }
    [scroll scrollRectToVisible:CGRectMake(0, 0.0f, 320.0f, App_SCREEN_
HEIGHT) animated:NO];
}
-(void)scrollViewDidScroll:(UIScrollView *)scrollView
{
    int page=scrollView.contentOffset.x / scrollView.frame.size.width;
    //设置页码
    pageControl.currentPage=page;
}
//返回 scrollView 中的第一个视图
-(UIView *)viewForZoomingInScrollView:(UIScrollView *)scrollView
{
    return [scrollView.subviews objectAtIndex:0];
}
```

```
//设置缩小最小值
- (void) scrollViewDidEndZooming: (UIScrollView  * ) scrollView  withView:
(UIView * )view atScale:(float)scale
{
    if(scrollView !=scroll){
        if(scale<1.0){
            [scrollView setZoomScale:1.0 animated:YES];
        }
    }
}
-(void)chickImage{
    [self dismissViewControllerAnimated:YES completion:nil];
}

-(void)didReceiveMemoryWarning
{
    [super didReceiveMemoryWarning];
    //Dispose of any resources that can be recreated.
}

@end
```

代码解析：

UIPageControl控件在程序中出现的比较频繁，尤其与UIScrollView配合显示大量数据时，会使用它来控制UIScrollView的翻页。在滚动ScrollView时可通过PageControll中的小白点来观察当前页面的位置，也可通过单击PageContrll中的小白点来滚动到指定的页面。

UIScrollView的常用属性如下：

contentOffSet——监控目前滚动的位置。

contentSize——滚动范围的大小。

contentInset——视图在scrollView中的位置。

delegate——设置代理。

directionalLockEnabled——指定控件是否只能在一个方向上滚动。

bounces——控制控件遇到边框是否反弹。

alwaysBounceVertical——控制垂直方向遇到边框是否反弹。

alwaysBounceHorizontal——控制水平方向遇到边框是否反弹。

pagingEnabled——控制控件是否整页翻动。

scrollEnabled——控制控件是否能滚动。

showsHorizontalScrollIndicator——控制是否显示水平方向的滚动条。

showsVerticalScrollIndicator——控制是否显示垂直方向的滚动条。

scrollIndicatorInsets——指定滚动条在scrollerView中的位置。

indicatorStyle——设定滚动条的样式。

decelerationRate——改变 scrollerView 的减速点位置。

tracking——监控当前目标是否正在被跟踪。

dragging——监控当前目标是否正在被拖曳。

decelerating——监控当前目标是否正在减速。

delaysContentTouches——控制视图是否延时调用开始滚动的方法。

canCancelContentTouches——控制控件是否接触取消 touch 的事件。

minimumZoomScale——缩小的最小比例。

maximumZoomScale——放大的最大比例。

zoomScale——设置变化比例。

bouncesZoom——控制缩放的时候是否会反弹。

zooming——判断控件的大小是否正在改变。

zoomBouncing——判断是否正在进行缩放反弹。

此外,UIScrollView 还有很多代理方法,一些常用的代理方法如下:

scrollViewDidScroll——滚动触发。

scrollViewWillBeginDragging——开始拖曳视图。

scrollViewDidEndDragging: willDecelerate——完成拖曳。

scrollViewWillBeginDecelerating——将开始降速。

scrollViewDidEndDecelerating——减速停止时执行,手触摸时执行。

scrollViewDidEndScrollingAnimation——滚动动画停止时执行,代码改变时触发。

viewForZoomingInScrollView——设置放大、缩小的视图。

在 loadView 方法中,我们创建了 UIScrollView 和 UIPageControl 的实例,并在 scrollViewDidScroll: 方法中将 UIScrollView 滚动页码赋值给 UIPageControl 的 currentPage 属性,实现小白点的移动。

要实现图片的放大、缩小,需要设置 UIScrollView 的 delegate 属性为当前 ViewController,并实现 UIScrollViewDelegate 协议中的 viewForZoomingInScrollView: 方法和 scrollViewDidEndZooming:withView:atScale:方法。

(5) 单击 ShopDetailViewController.m,导入头文件 ImagesViewController.h,代码如下:

```
#import "ImagesViewController.h"
```

(6) 重写 imageChick 方法。为方便讲解,将一个装有两个相同图片地址的数组传递给 ImagesViewController,模拟实现多张图片的切换效果。代码如下:

```
-(void) imageChick{
    ImagesViewController *vc=[[ImagesViewController alloc]init];
    vc.arrayImage=@[_shopItem.shopImage,_shopItem.shopImage];
    [self presentViewController:vc animated:YES completion:nil];
}
```

完成后单击 Xcode 左上角的▶按钮,运行程序。

6.5 基础知识与技能回顾

本章主要介绍了 UITableView 的用法、自定义 Cell 的实现、手势的添加以及图片的缩放与分页。

在 iOS 开发中,UITableView 可以说是使用最广泛的控件,我们平时使用的软件中到处都可以看到它的影子,类似于微信、QQ、新浪微博等软件基本上随处都是 UITableView。

UITableView 中的数据只有行的概念,没有列的概念,因为在手机操作系统中显示多列是不利于操作的。UITableView 中每行数据都是一个 UITableViewCell,在这个控件中为了显示更多的信息,iOS 已经在其内部设置好了多个子控件供开发者使用。

由于 iOS 是遵循 MVC 模式设计的,因此很多操作都是通过代理和外界沟通的,但对于数据源控件除了代理还有一个数据源属性,通过它和外界进行数据交互。对于 UITableView,设置完 dataSource 还需要实现 UITableViewDataSource 协议。

练　　习

项目 1

功能描述: 用 UITableView + 自定义 UITableViewCell 展示 4 个商户。UITableViewCell 需包括商户图片、商户名称。

项目 2

功能描述: 请求服务端,将服务端数据解析并显示在项目 1 创建的界面上。

第7章 chapter 7

支持用户基于LBS的应用

本章主要介绍地图(MapKit框架)和位置(Core Location框架),以及手机摇一摇的功能。

MapKit框架主要提供4个功能:显示地图、CLLocation和地址间的转化、支持在地图上做标记(如标记北京天安门)和将位置解析成地址。

Core Location框架为设备提供定位功能以确定和报告其位置(Location Services)。即使一个没有GPS或蜂窝能力的设备也可能通过扫描附近的WiFi设备,并同网上的数据库比对来实现位置服务。

摇一摇是一个非常有趣的功能,设备可以检测用户是否摇晃了手机,从而做出响应。

7.1 用户定位

在手机应用程序中,有时需要获取用户所在的当前位置,从而为其提供在该位置附近的信息。如果知道某个经纬度,同样能在地图上标记出该经纬度的位置,如图7.1所示。

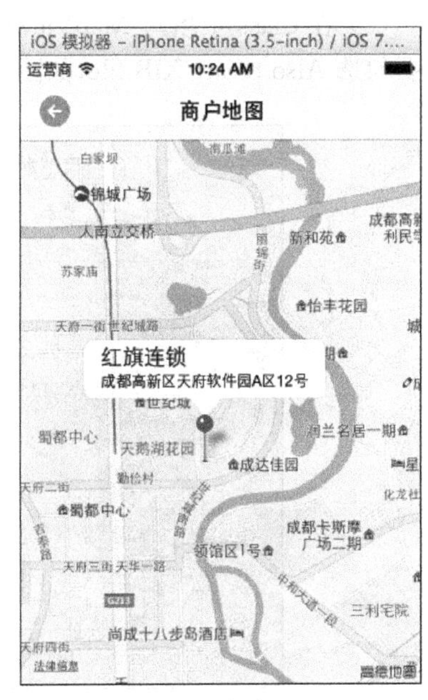

图7.1 商户位置

7.1.1 LBS与常见第三方地图

LBS(Location Based Services)又称位置服务。LBS是将移动通信网络和卫星定位系统结合在一起的一种增值业务,它通过一组定位技术获得移动终端的位置信息(如经纬度坐标数据),并提供给移动用户本人或其他通信系统,实现各种与位置相关的业务。实质上LBS是一种概念较为宽泛的与空间位置有关的新型服务业务。

第三方地图是指相对与iOS自带地图的地图服务。用户可以使用第三方地图服务查询街道、商场、楼盘的地理位置,也可以找到离用户最近的餐馆、学校、银行、公园等。目前iOS平台上主流的第三方地图有百度地图、SOSO地图等。

7.1.2 在地图上找到自己

要想在地图上显示自己的位置，首先需要获取当前位置的经纬度，然后标记这个经纬度并显示在地图上。

1. 添加地图页面

(1)选中 Sections 文件夹，按下 command＋N，选择 Cocoa Touch 下的 Objective-C class，如图 7.2 所示。

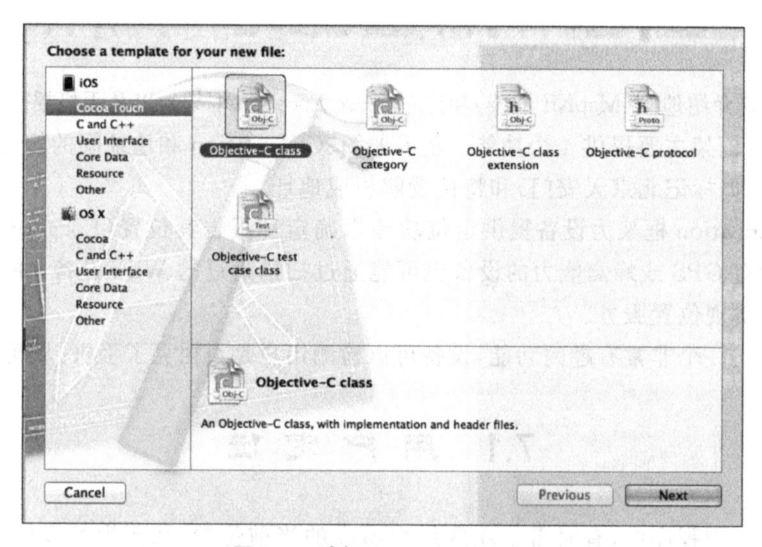

图 7.2 选择 Objective-C class

(2) 在 Class 中填写 MapViewController，在 SubClass of 中填写 FatherViewController，不要勾选 Also create XIB file，如图 7.3 所示。

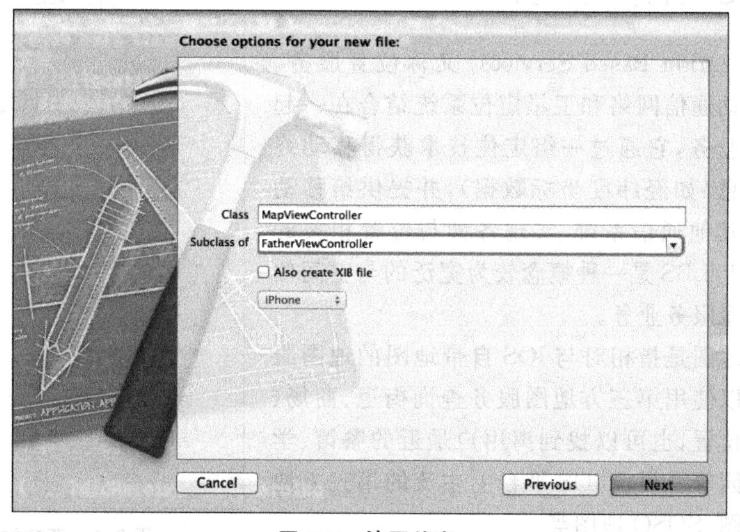

图 7.3 填写信息

(3) 依次单击工程→TARGETS(MyDemo)→Build Phase→Link Binary With Libraries,如图 7.4 所示。

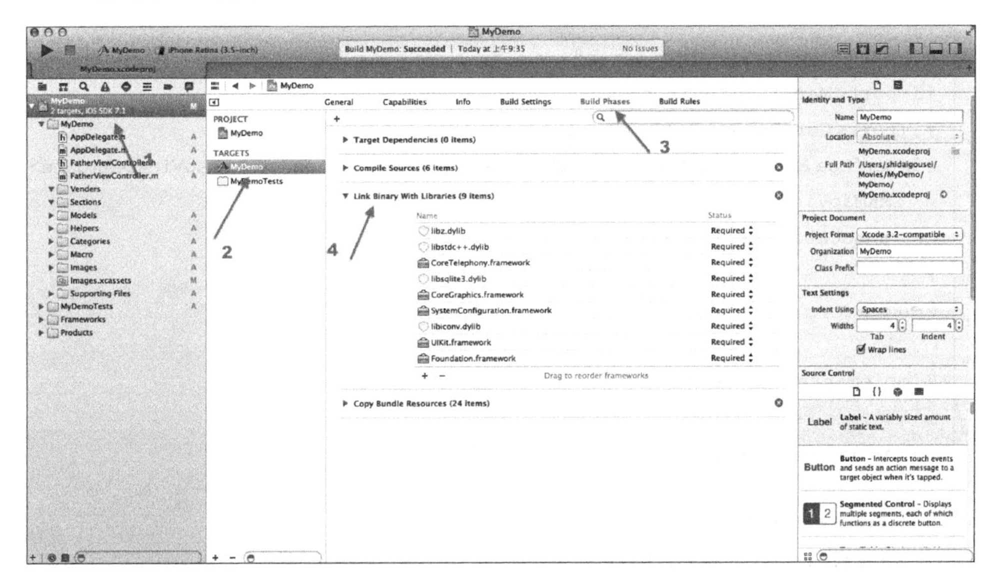

图 7.4 选择 Link Binary With Libraries

(4) 单击"+"号,分别添加 MapKit. framework 和 CoreLocation. frame,如图 7.5 所示。

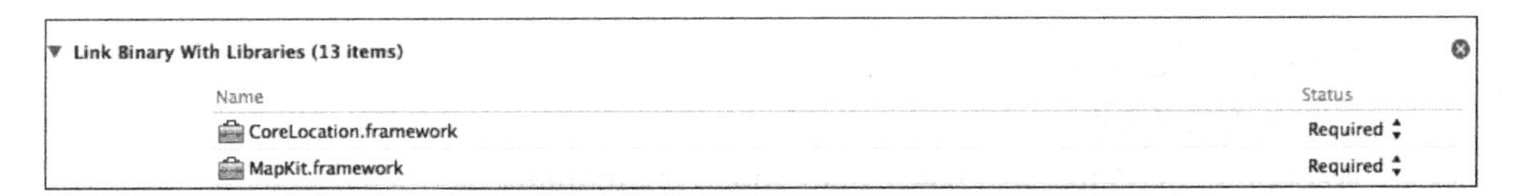

图 7.5 添加依赖框架

(5) 单击 MapViewController. h,设置 4 个属性用来接收商户名、商户地址、商户纬度、商户经度,更改代码如下:

```
#import "FatherViewController.h"
@interface MapViewController: FatherViewController

@property (nonatomic,copy)NSString * shopName;
@property (nonatomic, copy)NSString * shopAddress;
@property (nonatomic, copy)NSString * shopLatitude;
@property (nonatomic, copy)NSString * shopLongitude;

@end
```

代码解析:

这里也可以创建一个 ShopItem 模型类来接收商户信息,由于地图视图只需用到商户名、商户地址、商户经纬度，因此直接创建 4 个字符串来接收信息。

（6）单击 MapViewController.m，更改代码如下：

```
#import "MapViewController.h"
#import<MapKit/MapKit.h>
#import<CoreLocation/CoreLocation.h>
@interface MapViewController ()<CLLocationManagerDelegate,MKMapViewDelegate>{
    MKMapView *myMapView;
    CLLocationManager* mylm;
}
@end

@implementation MapViewController

- (void)viewDidLoad
{
    [super viewDidLoad];
    self.title=@"商户地图";
    [self createLeftBtnWithFrame:CGRectMake(0.0f, 0.0f, 24.0f, 24.0f) Image:
[UIImage imageNamed:@"backButtonNormal"]];
    //创建一个 CLLocationManager 实例
    mylm=[[CLLocationManager alloc] init];
    //设置委托
    mylm.delegate=self;
    //设置为最好精度
    mylm.desiredAccuracy=kCLLocationAccuracyBest;
    //设置距离筛选器 distanceFilter,下面表示设备至少移动 10 米,才通知委托更新
    mylm.distanceFilter=10;
    //启动定位
    [mylm startUpdatingLocation];

    //创建地图
    myMapView=[[MKMapView alloc] initWithFrame:self.view.bounds];
    myMapView.delegate=self;
    //显示用户当前位置
    myMapView.showsUserLocation=YES;
    //设置地图类型为标准地图模式
    myMapView.mapType=MKMapTypeStandard;
    [self.view addSubview:myMapView];

    CLLocationCoordinate2D  coordinate  =  CLLocationCoordinate2DMake  ([_
shopLatitude floatValue], [_shopLongitude floatValue]);
```

```
    //实例化一个地图标记
    MKPointAnnotation * annotation=[[MKPointAnnotation alloc] init];
    annotation.coordinate=coordinate;
    annotation.title=_shopName;
    annotation.subtitle=_shopAddress;
    [myMapView addAnnotation:annotation];
}
//定位回调函数
- (void) locationManager: (CLLocationManager * )manager didUpdateToLocation:
(CLLocation * )newLocation fromLocation:(CLLocation * )oldLocation{
    //确定比例
    MKCoordinateSpan span=MKCoordinateSpanMake(0.01, 0.01);
    //根据当前位置和比例进行定位
    MKCoordinateRegion   region   =   MKCoordinateRegionMake   ( newLocation.
coordinate, span);
    [myMapView setRegion:region animated:YES];
    //停止定位
    [manager stopUpdatingLocation];
}
//定位失败
- (void) locationManager: (CLLocationManager  * ) manager didFailWithError:
(NSError * )error{
    [manager stopUpdatingLocation];
}
- (void)leftBarChick{
    [self dismissViewControllerAnimated:YES completion:nil];
}
- (void)didReceiveMemoryWarning
{
    [super didReceiveMemoryWarning];
    //Dispose of any resources that can be recreated.
}
    @end
```

代码解析：

首先创建一个 CLLocationManager 对象，用来获取当前设备的位置。因为对定位设备的轮询是很耗电的，所以最好只在非常必要的前提下启动。

定位有 3 种方式：

(1) GPS——最精确的定位方式。

(2) 蜂窝基站三角定位——这种定位在信号基站比较密集的城市比较准确。

(3) WiFi——这种方式是通过网络运营商的数据库得到数据，在 3 种定位中最不精确。

然后创建一个 MKMapView 的实例，初始化地图，用来显示设备当前位置。

最后创建一个 MKPointAnnotation 实例，用来显示商户的位置。这里需要为 MKPointAnnotation 实例设置坐标，然后将其添加到地图上。

2. 为地图页面增加入口

(1) 单击 ShopDetailViewController.m，导入头文件，代码如下：

```
#import "MapViewController.h"
```

(2) 重写 goMap:方法，代码如下：

```
-(IBAction)goMapVC:(id)sender {
    MapViewController *vc=[[MapViewController alloc]init];
    vc.shopName=_shopItem.shopName;
    vc.shopAddress=_shopItem.shopAddress;
    vc.shopLatitude=_shopItem.shopLatitude;
    vc.shopLongitude=_shopItem.shopLongitude;
    UINavigationController *nav = [[UINavigationController alloc] initWith
RootViewController:vc];
    [self presentViewController:nav animated:YES completion:nil];
}
```

代码解析：

在 goMapVC：方法中，将商户名、商户地址、商户经纬度分别传递给地图页面的对应属性。

(3) 单击左上角的▶按钮运行程序。注意，模拟器默认不能定位，不过可以手动给模拟器设置一个经纬度。选中模拟器，依次选择：调试→位置→自定义位置，输入经纬度后，单击“好”就能实现定位了。

7.2 摇 一 摇

对于摇一摇，很多手机用户都会想到腾讯旗下的微信，其实摇一摇并非微信独有的功能。基于手机自身的重力感应，一些拼图软件、安全软件等也具有摇一摇的功能。

本节将介绍通过摇一摇查找周边商户的功能，大致原理如下：

(1) 摇晃手机时，传感器如果检测到手机正在摇动，就会向附近的手机基站发出请求。

(2) 附近手机基站接收到请求并返回基站标号。

(3) 将当前位置经纬度发送到服务端后台。

(4) 服务端后台根据经纬度计算周边商户，并返回给应用。

(5) 应用解析服务端返回的数据并显示。

本项目摇一摇的界面如图 7.6 所示。

图 7.6 摇一摇的界面

7.2.1 客户端代码开发

1. 添加摇一摇的页面

(1) 选中 Sections 文件夹，按下 command＋N，选择 Cocoa Touch 下的 Objective-C class，如图 7.7 所示。

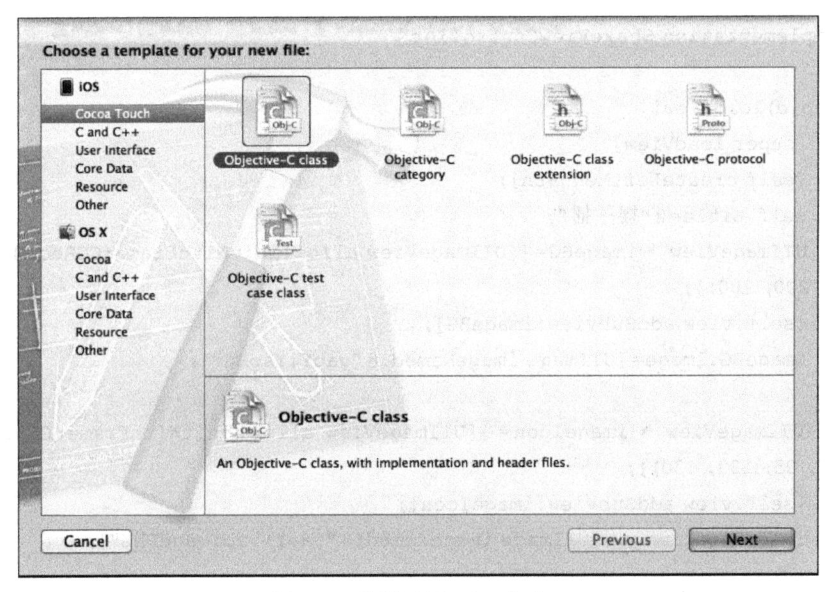

图 7.7 选择 Objective-C class

（2）在 Class 中填写 ShakeViewController，在 SubClass of 中填写 FatherView Controller，不要勾选 With XIB for user interface，如图 7.8 所示。

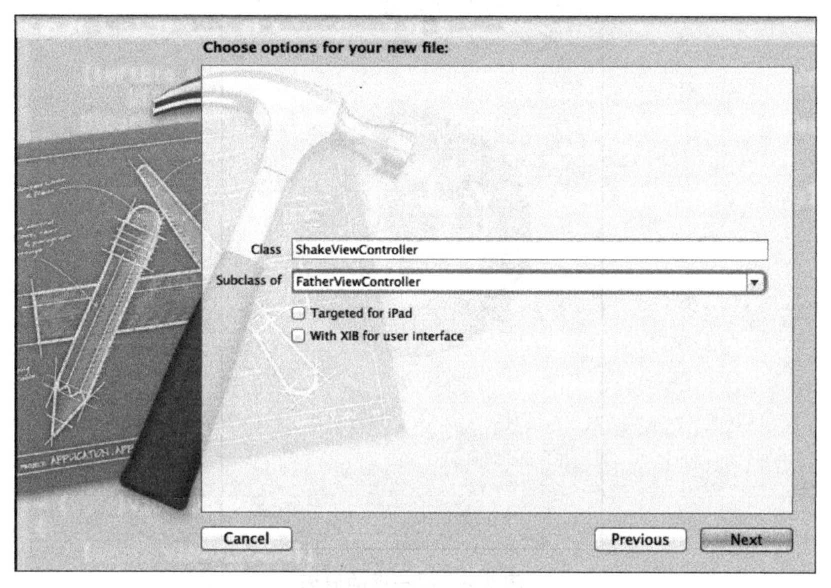

图 7.8 填写信息

（3）单击 ShakeViewController.m，更改代码如下：

```
#import "ShakeViewController.h"
@interface ShakeViewController ()

@end

@implementation ShakeViewController

- (void)loadView{
    [super loadView];
    [self createLeftMenuBtn];
    self.title=@"摇一摇";
    UIImageView * imageBG= [[UIImageView alloc]initWithFrame:CGRectMake(60,
60, 200, 200)];
    [self.view addSubview:imageBG];
    imageBG.image= [UIImage imageNamed:@"yaoYiYaoBG"];

    UIImageView * imageIcon= [[UIImageView alloc]initWithFrame:CGRectMake
(95, 95, 130, 130)];
    [self.view addSubview:imageIcon];
    imageIcon.image= [UIImage imageNamed:@"yaoYiYaoImagePNG"];
}
```

```
- (void)viewDidAppear:(BOOL)animated{
    [super viewDidAppear:animated];
    [self becomeFirstResponder];
}
- (void)viewWillDisappear:(BOOL)animated{
    [self resignFirstResponder];
    [super viewWillDisappear:animated];
}

- (void)viewDidLoad{
    [super viewDidLoad];
}
//这里需要返回 YES
- (BOOL)canBecomeFirstResponder{
    return YES;
}
//摇动开始
- (void)motionBegan:(UIEventSubtype)motion withEvent:(UIEvent *)event{
}
//摇动结束
- (void)motionEnded:(UIEventSubtype)motion withEvent:(UIEvent *)event{
    if(motion==UIEventSubtypeMotionShake){
    //这里面插入事件代码
    }
}
//摇动取消
- (void)motionCancelled:(UIEventSubtype)motion withEvent:(UIEvent *)event{
}

- (void)didReceiveMemoryWarning{
    [super didReceiveMemoryWarning];
    //Dispose of any resources that can be recreated.
}
@end
```

代码解析：

首先需要在 viewDidAppear 方法中调用 canBecomeFirstResponder 方法，并重写 canBecomeFirstResponde，返回 YES，让当前 ViewController 支持摇动。

然后分别实现 motionBegan：withEvent、motionEnded：event、motionCancelled：withEvent 方法，编写对应的动作。

2. 为摇一摇页面添加入口

(1) 单击 LeftViewController. xib，拖放一个 UIButton 控件到 xib，双击命名为"摇一摇"，如图 7.9 所示。

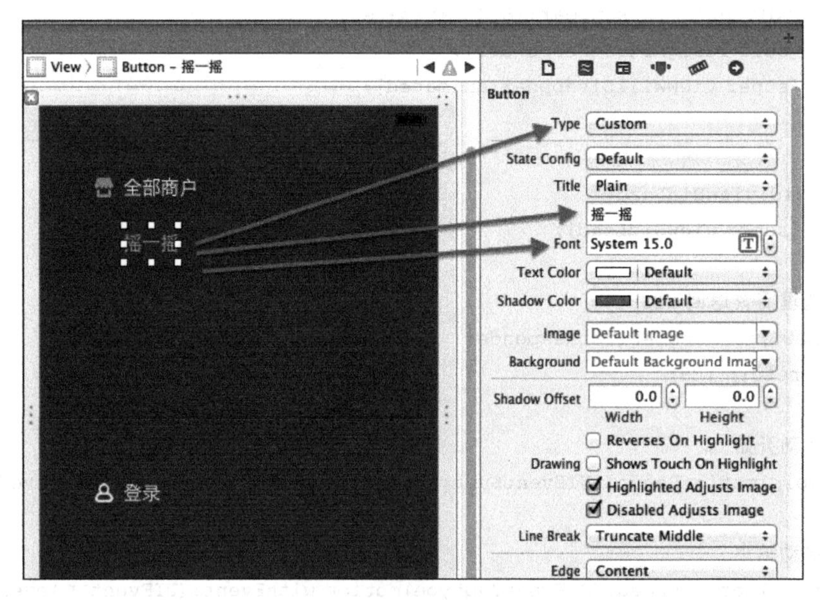

图 7.9　拖放 UIButton 控件

(2) 拖放一个 UIImageView 控件，Image 属性为 iconShakeNormal. png，如图 7.10 所示。

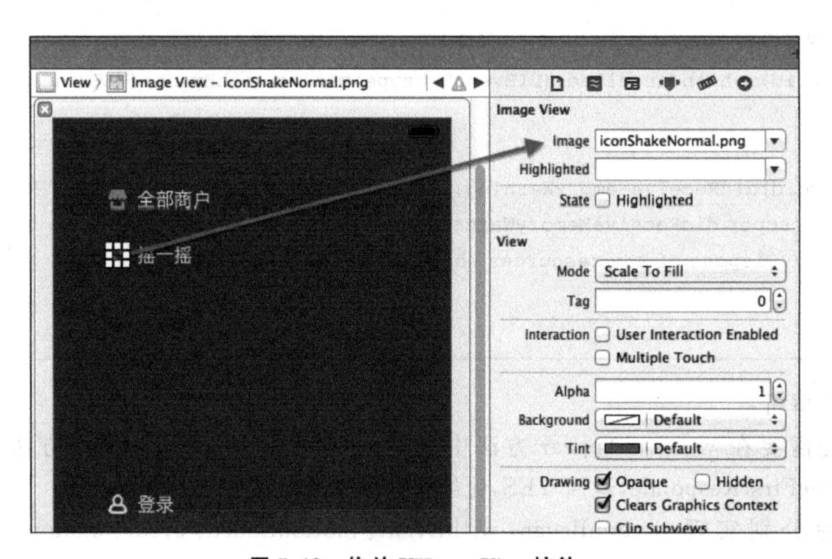

图 7.10　拖放 UIImageView 控件

(3) 为摇一摇按钮关联 Touch Up Inside 事件，并命名为 goShake，如图 7.11 所示。

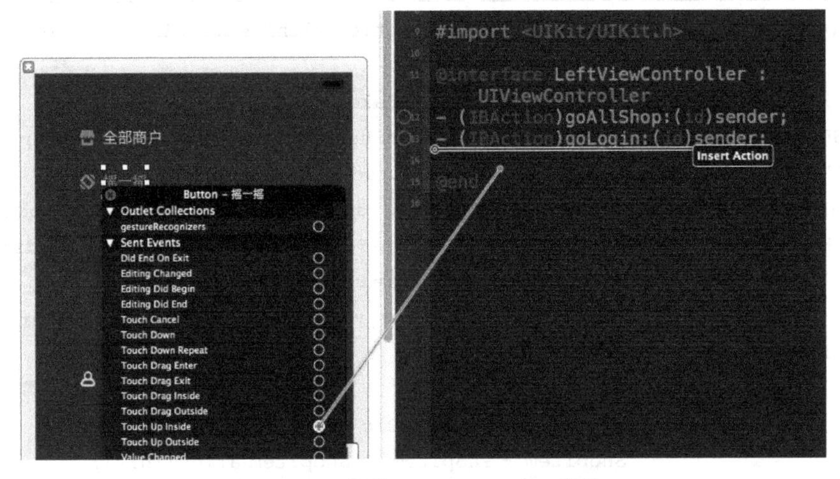

图 7.11 关联 Touch Up Inside 事件

(4) 单击 LeftViewController.m，导入头文件，代码如下：

```
#import "ShakeViewController.h"
```

(5) 重写 goShake:方法，代码如下：

```
-(IBAction)goShake:(id)sender {
    ShakeViewController *vc=[[ShakeViewController alloc]init];
    UINavigationController * navigationController = self. menuContainerView
Controller.centerViewController;
    NSArray *controllers=[NSArray arrayWithObject:vc];
    navigationController.viewControllers=controllers;
    [self.menuContainerViewController setMenuState:MFSideMenuStateClosed];
}
```

代码解析：

单击 goShake:方法后，程序将摇一摇页面设置为侧滑的中心页，并关闭左侧滑菜单页。

7.2.2 客户端与服务端交互

(1) 单击 ShakeViewController.m，导入头文件，代码如下：

```
#import "AFHelper.h"
#import "AllShopViewController.h"
#import "ShopItem.h"
```

(2) 为方便讲解，这里用第 6 章创建的全部商户页来展示摇一摇的结果。重写 motionEnd:withEvent:方法，代码如下：

```
- (void)motionEnded: (UIEventSubtype)motion withEvent: (UIEvent * )event
{
    if(motion==UIEventSubtypeMotionShake){
NSDictionary * dic=@ {@"act":@"searchShop",@"pageSize":@"20",@"page":@
"1",@"longitude":@"104.074974",@"latitude":@"30.560876"};
[AFHelper downDataWithDictionary: dic andBaseURLStr: @" http://localhost:
8080/meServer/" andPostPath:@"shop.php?" success:^(NSDictionary * dic){
            if([dic[@"flag"]intValue]==1){
                AllShopViewController * vc= [[AllShopViewController alloc]
init];
                vc.titile=@"商户列表";
                for(NSDictionary * tmpDic in dic[@"vendorList"]){
                    ShopItem * shopItem= [[ShopItem alloc]init];
                    shopItem.shopId=tmpDic[@"shopId"];
                    shopItem.shopAddress=tmpDic[@"shopAddress"];
                    shopItem.shopDesc=tmpDic[@"shopDesc"];
                    shopItem.shopImage=tmpDic[@"shopImage"];
                    shopItem.shopLatitude=tmpDic[@"shopLatitude"];
                    shopItem.shopLongitude=tmpDic[@"shopLongitude"];
                    shopItem.shopName=tmpDic[@"shopName"];
                    shopItem.shopPhone=tmpDic[@"shopPhone"];
                    shopItem.shopSpend=tmpDic[@"shopSpend"];
                    shopItem.shopTime=tmpDic[@"shopTime"];
                    shopItem.shopType=tmpDic[@"shopType"];
                    shopItem.shopDistance=tmpDic[@"distance"];
                    [vc.dataArray addObject:shopItem];
                }
                UINavigationController * nav = [[ UINavigationController
alloc]initWithRootViewController:vc];
                [self presentViewController:nav animated:YES completion:nil];
            }else{
                UIAlertView * alert= [[UIAlertView alloc]initWithTitle:@"提
示" message: dic [@" msg "] delegate: self cancelButtonTitle: @" 好 的 "
otherButtonTitles: nil ,nil];
                [alert show];
            }
        }];
    }
}
```

代码解析：

motionEnd:withEvent:方法是摇动结束后触发的方法，程序将参数组合并请求 http://localhost:8080/meServer/shop.php 接口。为方便讲解，这里将经纬度固定填写。实际开发中，经纬度应该填写用户定位后的经纬度。

从服务端获取到数据后，程序将商户模型类添加到 AllShopViewController 的 dataArray 数组中。

(3) 之前全部商户页在加载前没有数据，需要初始化后主动发送请求到服务端获取数据；而现在全部商户页作为搜索结果展示，数据在初始化后就传递给了其 dataArray 属性，因此不需要再向服务端发送请求获取数据。增加一个判断，根据 dataArray 的 count 数量来判断是否需要请求商户数据。单击 AllShopViewController.m，添加判断（全部商户页还是商户列表页），代码如下：

```
- (void)viewDidLoad
{
    [super viewDidLoad];
    if(_dataArray.count<1){
        [self createLeftMenuBtn];
        [self getData];
    }else{
        [self createLeftBtnWithFrame:CGRectMake (0.0f, 0.0f, 24.0f, 24.0f)
Image:[UIImage imageNamed:@"backButtonNormal"]];
        [_table reloadData];
    }
}
- (void)leftBarChick{
    [self dismissViewControllerAnimated:YES completion:nil];
}
```

代码解析：

如果 AllShopViewController 作为全部商户页，则 dataArray 为空，需要请求服务端获取数据并创建左上角默认按钮，用来打开左侧滑页。如果作为搜索结果展示页，dataArray 已有数据，则直接填充到 tableView 并创建左上角的自定义按钮，用来返回上一个页面。

7.3 基础知识与技能回顾

本章介绍了如何使用定位功能以及在地图上显示位置。基于 LBS 可以开发许多重要的功能，比如交友、团购等。

本章还介绍了一个非常有趣的功能：摇一摇。通过编写简单的代码，就能检测用户摇晃手机的行为，编写相应的行为代码。比如，可以实现摇一摇抽奖、摇一摇换歌曲等功能。

练　习

项目 1

功能描述：在地图上显示一个经度：104.075181、纬度：30.552728 的位置。

项目 2

功能描述：使用摇一摇的功能，查找周边的商户信息。

第8章 chapter 8

让用户搜索

由于手机屏幕较小，因此手机一次展现的内容非常有限，而用户常常希望需要的信息能够迅速、直观地展现出来。结合项目实例，本章介绍3种搜索方式：

- 按关键字搜索。
- 按类别搜索。
- 按地理位置搜索。

8.1 服务端接口的准备

本节将用到搜索商户接口，下面给出接口的详细信息。

接口地址：http://localhost:8080/meServer/shop.php?

调用方式：Post

返回数据格式：Json

请求参数及说明如表8.1所示。

表8.1 搜索商户接口请求参数

请求参数	必选	类型	说明
act	Y	string	searchShop
pageSize	Y	string	每页数量
page	Y	string	页号
latitude	N	string	纬度
longitude	N	string	经度
type	N	string	商户类型：传数字1(餐饮美食)、2(休闲娱乐)、3(全部类型)
range	N	string	商户区域：传数字1(500米)、2(1000米)、3(全部区域)

返回字段及说明如表8.2所示。

表8.2 搜索商户接口返回字段

返回字段	字段类型	字段说明	返回字段	字段类型	字段说明
flag	string	0：失败，1：成功	vendorList	JsonArray	商户信息数组
msg	string	信息说明	shopAddress	string	地址

续表

返回字段	字段类型	字段说明	返回字段	字段类型	字段说明
shopDesc	string	描述	shopPhone	string	商户电话
shopId	string	商户 id	shopSpend	string	商户人均花费
shopImage	string	商户图片	shopTime	string	商户营业时间
shopLatitude	string	商户纬度	shopType	string	商户类型
shopLongitude	string	商户经度	distance	string	商户距离
shopName	string	商户名字			

8.2 常用搜索方式与应用开发

搜索界面最终效果如图 8.1 和图 8.2 所示。

图 8.1 搜索界面

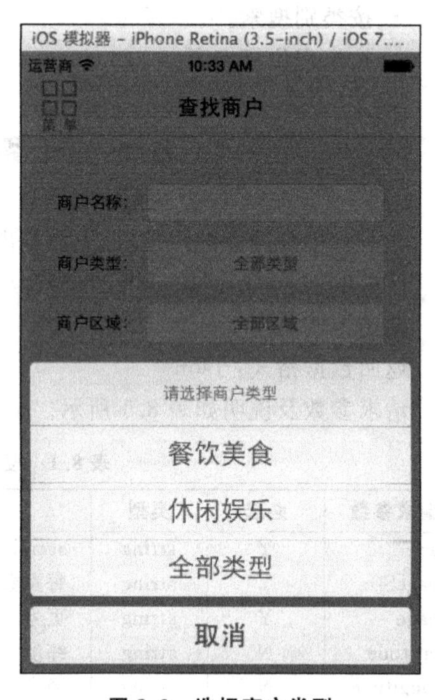

图 8.2 选择商户类型

8.2.1 客户端代码开发

1. 添加搜索页面

(1) 按下 command+N,选择 iOS 下的 Objective-C class,如图 8.3 所示。

(2) 在 Class 中填写 SearchViewController,在 SubClass of 中填写 FatherViewController,勾选 Also create XIB file,如图 8.4 所示。

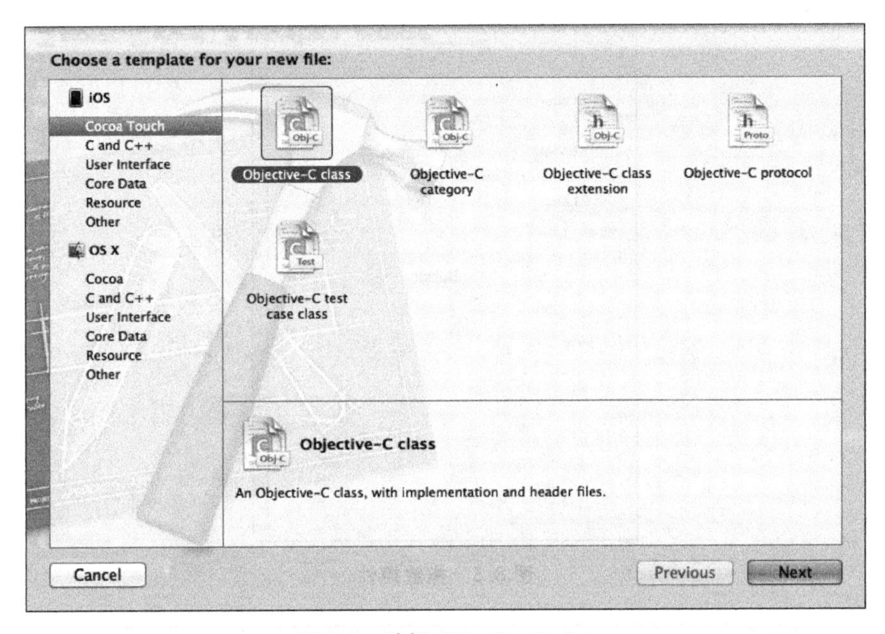

图 8.3 选择 Objective-C class

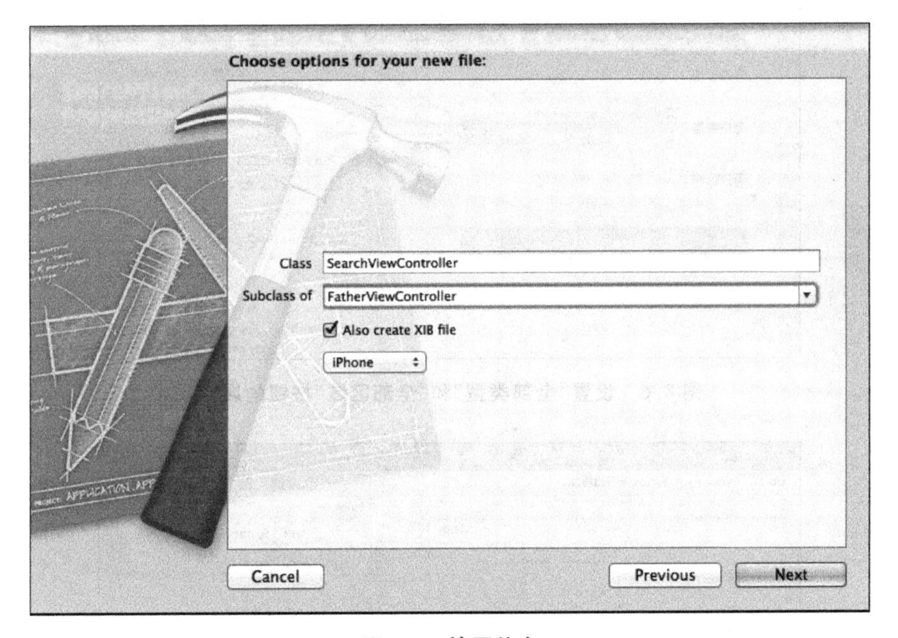

图 8.4 填写信息

(3) 拖放 1 个 UITextField、3 个 UILabel、3 个 UIButton 控件到 xib 上，如图 8.5 所示。

(4) 分别双击命名 UILabel 和 UIButton，并设置 UIButton 背景图片，如图 8.6 和图 8.7 所示。

(5) 关联控件到 SearchViewController.h 文件，如图 8.8 所示。

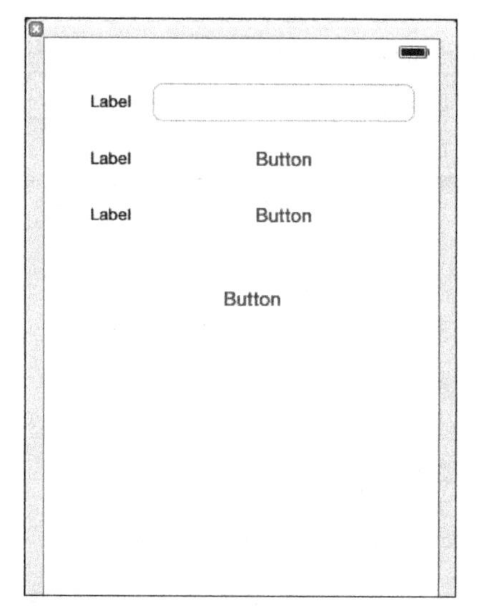

图 8.5　拖放控件

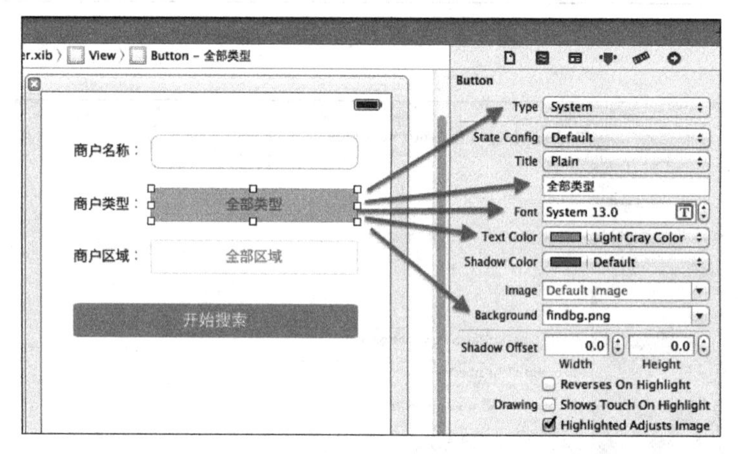

图 8.6　设置“全部类型”和“全部区域”按钮的属性

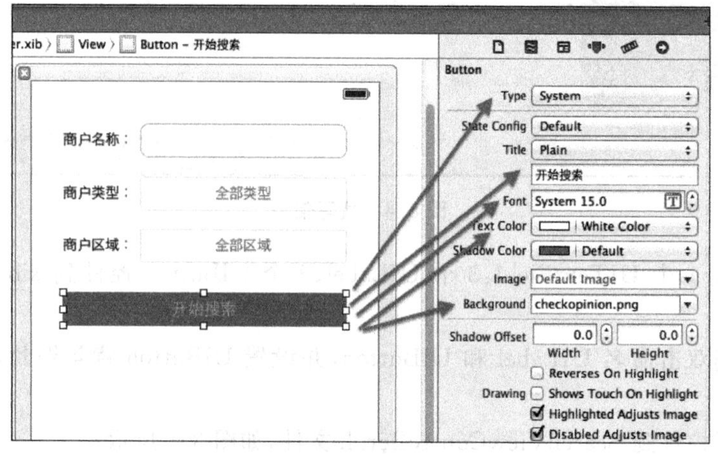

图 8.7　设置“开始搜索”按钮的属性

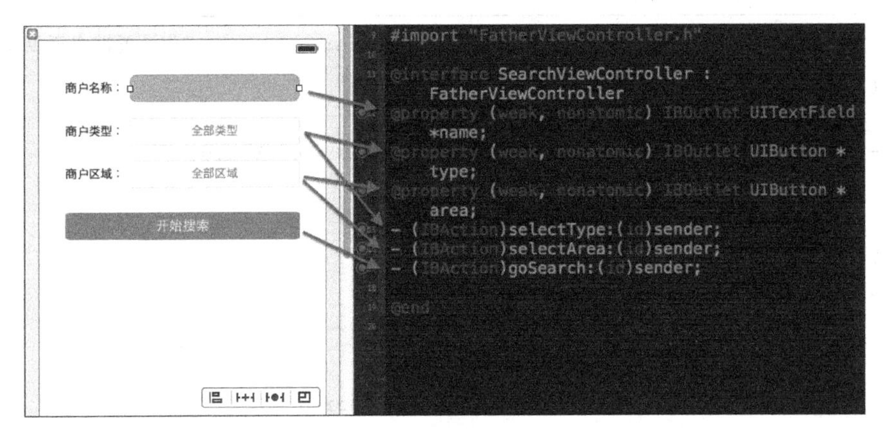

图 8.8　关联控件

(6) 单击 SearchViewController.m,代码如下:

```
#import "SearchViewController.h"
@interface SearchViewController ()<UITextFieldDelegate,UIActionSheetDelegate>

@property (assign)NSUInteger searchType;            //搜索类型
@property (assign)NSUInteger searchRange;           //搜索范围

@end

@implementation SearchViewController

- (id) initWithNibName: (NSString  * ) nibNameOrNil  bundle: (NSBundle  * )
nibBundleOrNil
{
    self= [super initWithNibName:nibNameOrNil bundle:nibBundleOrNil];
    if (self) {
        //Custom initialization
    }
    return self;
}

- (void)viewDidLoad
{
    [super viewDidLoad];
    [self createLeftMenuBtn];
    self.title=@"查找商户";
    _searchType=3;                                  //初始化搜索类型为全部类型
    _searchRange=3;                                 //初始化搜索范围为全部范围
    self.name.delegate=self;
```

```
    self.name.returnKeyType=UIReturnKeyDone;                //设置键盘返回键类型
}
//单击键盘右下角的按钮,让键盘消失
-(BOOL)textFieldShouldReturn:(UITextField *)textField{
    [textField resignFirstResponder];
    return YES;
}
-(void)didReceiveMemoryWarning
{
    [super didReceiveMemoryWarning];
    //Dispose of any resources that can be recreated.
}

-(IBAction)selectType:(id)sender {
    UIActionSheet * sheet=[[UIActionSheet alloc]initWithTitle:@"请选择商户
类型" delegate:self cancelButtonTitle:@"取消" destructiveButtonTitle:nil
otherButtonTitles:@"餐饮美食",@"休闲娱乐",@"全部类型", nil];
    sheet.tag=100;
    [sheet showInView:self.view];
}

-(IBAction)selectArea:(id)sender {
    UIActionSheet * sheet=[[UIActionSheet alloc]initWithTitle:@"请选择商户
区域" delegate:self cancelButtonTitle:@"取消" destructiveButtonTitle:nil
otherButtonTitles:@"附近500米",@"附近1000米",@"全部区域", nil];
    sheet.tag=101;
    [sheet showInView:self.view];
}
-(void)actionSheet:(UIActionSheet *)actionSheet clickedButtonAtIndex:
(NSInteger)buttonIndex{
    if(actionSheet.tag==100){
        _searchType=buttonIndex+1;
        switch (buttonIndex) {
            case 0:
                [_type setTitle:@"餐饮美食" forState:UIControlStateNormal];
                break;
            case 1:
                [_type setTitle:@"休闲娱乐" forState:UIControlStateNormal];
                break;
            case 2:
                [_type setTitle:@"全部类型" forState:UIControlStateNormal];
                break;
            default:
```

```
                break;
            }
    }else{
        _searchRange=buttonIndex+1;
        switch (buttonIndex) {
            case 0:
                [_area setTitle:@"附近 500 米" forState:UIControlStateNormal];
                break;
            case 1:
                [_area setTitle:@"附近 1000 米" forState:UIControlStateNormal];
                break;
            case 2:
                [_area setTitle:@"全部区域" forState:UIControlStateNormal];
                break;
            default:
                break;
        }
    }
}

-(IBAction)goSearch:(id)sender {
}
@end
```

代码解析：

这里用到了 UIActionSheet。UIActionSheet 是在 iOS 弹出的选择按钮项，可以添加多项，并为每项添加单击事件。UIActionSheet 的使用方法如下：

为了让控制器类充当操作表的委托，控制器类需要遵从协议：UIActionSheetDelegate。

生成 UIActionSheet 并显示，代码如下：

```
UIActionSheet * sheet=[[UIActionSheet alloc]initWithTitle:@"请选择商户类型"
delegate: self  cancelButtonTitle: @" 取 消 "  destructiveButtonTitle: nil
otherButtonTitles:@"餐饮美食",@"休闲娱乐",@"全部类型", nil];
[sheet showInView:self.view];
```

单击按钮触发事件，代码如下：

```
-(void) actionSheet: (UIActionSheet  * ) actionSheet clickedButtonAtIndex:
(NSInteger)buttonIndex{
    switch (buttonIndex) {
            case 0:
                break;
```

```
            case 1:
                break;
            case 2:
                break;
            default:
                break;
        }
}
```

searchType 和 searchRange 分别用来记录用户选择的搜索商户类型和搜索商户区域，默认都为全部选中。

2. 添加搜索入口

(1) 单击 LeftViewController.xib，拖放 1 个 UIButton 到 xib，双击命名为"查找商户"，如图 8.9 所示。

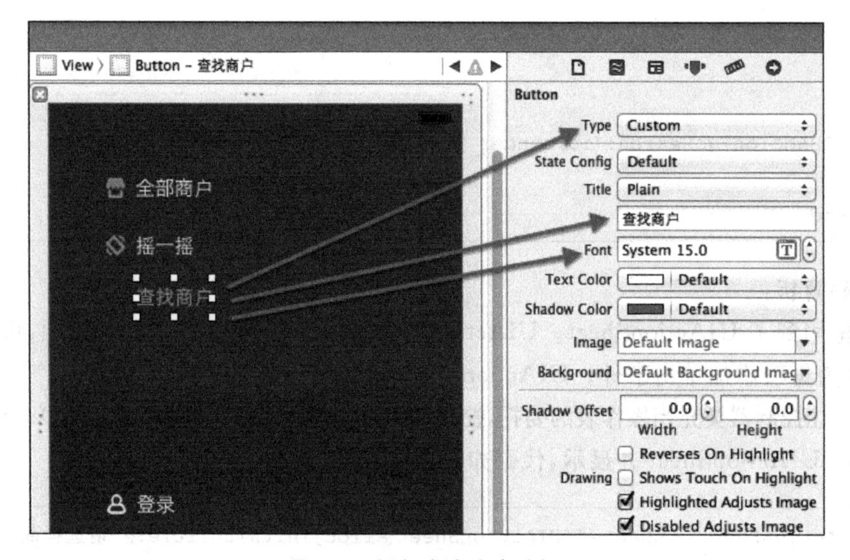

图 8.9　添加查找商户按钮

(2) 拖放 1 个 UIImageView 到 xib，设置 Image 属性为 iconSearchNormal.png，如图 8.10 所示。

(3) 为"查找商户"按钮绑定 Touch Up Inside 事件，并命名为 goSearch，如图 8.11 所示。

(4) 单击 LeftViewController.m 文件，导入头文件，代码如下：

```
#import "SearchViewController.h"
```

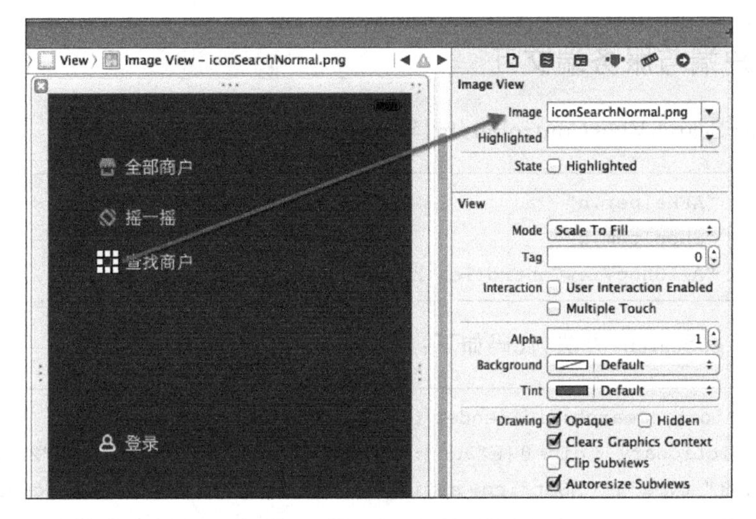

图 8.10 设置图片属性

图 8.11 关联事件

(5) 重写 goSearch:方法，代码如下：

```
-(IBAction)goSearch:(id)sender {
    SearchViewController *vc=[[SearchViewController alloc]init];
    UINavigationController * navigationController = self. menuContainerView
Controller.centerViewController;
    NSArray *controllers=[NSArray arrayWithObject:vc];
    navigationController.viewControllers=controllers;
    [self.menuContainerViewController setMenuState:MFSideMenuStateClosed];
}
```

代码解析：

单击 goShake:方法后，程序将搜索页面设置为侧滑的中心页，并关闭左侧滑菜单页。

8.2.2 客户端与服务端交互

（1）单击 SearchViewController.m，导入头文件，代码如下：

```
#import "AFHelper.h"
#import "ShopItem.h"
#import "AllShopViewController.h"
```

（2）重写 goSearch:方法，代码如下：

```
-(IBAction)goSearch:(id)sender {
    NSDictionary *dic=@{@"act":@"searchShop",@"pageSize":@"20",@"page":
@"1", @" type": [NSString stringWithFormat: @"% lu", (unsigned long) _
searchType],@"range":[NSString stringWithFormat:@"% lu", (unsigned long) _
searchRange], @" keyword": _name. text, @" longitude": @" 104. 074974", @
"latitude":@"30.560876"};
    [AFHelper downDataWithDictionary:dic andBaseURLStr:@"http://localhost:
8080/meServer/" andPostPath:@"shop.php?" success:^(NSDictionary *dic){
        if([dic[@"flag"]intValue]==1){
            AllShopViewController *vc=[[AllShopViewController alloc]init];
            for(NSDictionary *tmpDic in dic[@"vendorList"]){
                shopItem *shopItem=[[ShopItem alloc]init];
                shopItem.shopId=tmpDic[@"shopId"];
                shopItem.shopAddress=tmpDic[@"shopAddress"];
                shopItem.shopDesc=tmpDic[@"shopDesc"];
                shopItem.shopImage=tmpDic[@"shopImage"];
                shopItem.shopLatitude=tmpDic[@"shopLatitude"];
                shopItem.shopLongitude=tmpDic[@"shopLongitude"];
                shopItem.shopName=tmpDic[@"shopName"];
                shopItem.shopPhone=tmpDic[@"shopPhone"];
                shopItem.shopSpend=tmpDic[@"shopSpend"];
                shopItem.shopTime=tmpDic[@"shopTime"];
                shopItem.shopType=tmpDic[@"shopType"];
                shopItem.shopDistance=tmpDic[@"distance"];
                [vc.dataArray addObject:shopItem];
            }
            UINavigationController *nav = [[UINavigationController alloc]
initWithRootViewController:vc];
            [self presentViewController:nav animated:YES completion:nil];
        }else{
            UIAlertView *alert = [[UIAlertView alloc] initWithTitle: @"提示"
message:dic[@"msg"] delegate:nil cancelButtonTitle:@"好的" otherButtonTitles:
nil, nil];
```

```
            [alert show];
        }

    }];
}
```

代码解析：

单击搜索按钮，程序将参数组合后请求 http://localhost:8080/meServer/shop.php 接口。为方便讲解，这里将经纬度固定填写。实际开发中，经纬度应该填写用户定位后的经纬度。如果服务端有商户信息返回，则将数据传递给 AllShopViewController 的 dataArray 属性并显示 AllShopViewController 页。

8.3 基础知识与技能回顾

本章介绍了 3 种常用的搜索方式：按关键字搜索、按类别搜索、按地理位置搜索。这 3 种方式的实现都比较简单，基本原理是手机端提交相应搜索请求到服务端，服务端根据搜索请求查找数据并返回给手机端，手机端再解析数据并显示。

练 习

项目 1

功能描述：实现对商户按类别搜索。

项目 2

功能描述：实现对商户按两个以上组合搜索。

第9章 chapter 9

与用户互动

互动就是参与。由于无线网络“永不间断”的特性，互动与参与的可能性、积极性和随机性都超过了以往的传统网络。

手机 App 提供了比以往媒介更丰富多彩的表现形式。相比传统媒介的文字输入，移动设备的触摸屏有更好的操作体验。

互动性 App 打开了人与人的互动通道。比如通过在内部加入分享功能，使正在使用 App 的用户可以将感兴趣的内容分享给其他人，相互交流心得，迅速传播，积累用户。

9.1 数据库的准备

第 2 章已介绍了如何搭建项目服务端并配置完成本项目所需的后台数据。下面简要介绍数据库的准备。

评价表用来存储评价信息，如图 9.1 所示。

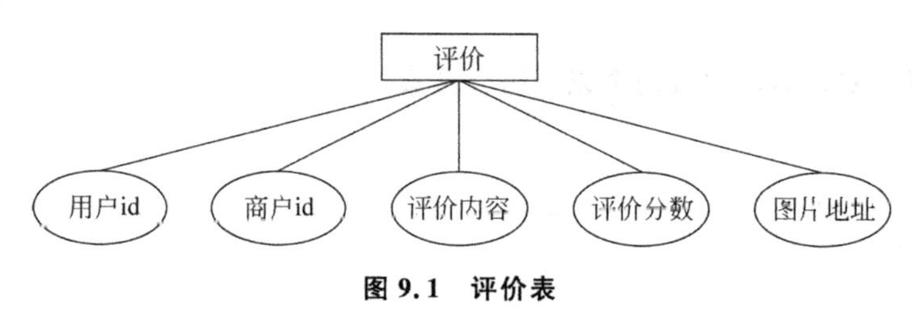

图 9.1 评价表

SQL 语句如下：

```
CREATE TABLE 'me_comment' (
  'appraise_id' int(10) unsigned NOT NULL AUTO_INCREMENT,
  'appraise_content' varchar(255) NOT NULL,
  'appraise_point' varchar(255) NOT NULL,
  'appraise_image_url' varchar(255) DEFAULT NULL,
  'user_name' varchar(255) NOT NULL,
  'shop_id' int(11) NOT NULL,
```

```
  'appraise_time' timestamp NOT NULL DEFAULT CURRENT_TIMESTAMP ON UPDATE
CURRENT_TIMESTAMP,
  PRIMARY KEY ('appraise_id')
) ENGINE=InnoDB AUTO_INCREMENT=37 DEFAULT CHARSET=utf8;
```

9.2 服务端接口的准备

本节将用到3个接口，其详细信息简介如下。

1. 显示商户评价接口

接口地址：http://localhost:8080/meServer/appraise.php?
调用方式：Post
返回数据格式：Json
请求参数及说明如表9.1所示。

表9.1 显示商户评价接口请求参数

请求参数	必选	类型	说明
act	Y	string	getAppraise
shopId	Y	string	商户id

返回字段及说明如表9.2所示。

表9.2 显示商户评价接口返回字段

返回字段	字段类型	字段说明	返回字段	字段类型	字段说明
flag	string	0：失败，1：成功	appraisePoint	string	分数
msg	string	信息说明	appraiseImage	string	图片地址
appraiseList	JsonArray	评价信息数组	userName	string	用户名
appraiseContent	string	内容			

2. 上传图片接口

接口地址：http://localhost:8080/meServer/image.php?
调用方式：Post
返回数据格式：Json
请求参数及说明如表9.3所示。

表9.3 上传图片接口请求参数

请求参数	必选	类型	说明
act	Y	string	postImage
file	Y	data	图片二进制

返回字段及说明如表 9.4 所示。

表 9.4 上传图片接口返回字段

返回字段	字段类型	字段说明
flag	string	0：失败，1：成功
msg	string	信息说明
image	string	图片地址

3. 上传商户评价接口

接口地址：http://localhost:8080/meServer/appraise.php?
调用方式：Post
返回数据格式：Json
请求参数及说明如表 9.5 所示。

表 9.5 上传商户评价接口请求参数

请求参数	必 选	类 型	说 明
act	Y	string	postAppraise
appraiseContent	Y	string	评价内容
appraisePoint	Y	string	评价分数
appraiseImageUrl	N	string	评价图片
userName	Y	string	用户名
shopId	Y	string	商户 id

返回字段及说明如表 9.6 所示。

表 9.6 上传商户评价接口返回字段

返回字段	字段类型	字段说明
flag	string	0：失败，1：成功
msg	string	信息说明

9.3 让用户参与评价

大家如果用 App Store 下载 App 就知道，App Store 有个显示评价页面，它是用户对某款 App 的使用评价。下载 App 后，可以通过评分、写评价的方式来发表对该 App 的看法。

本项目创建后的显示评价页面和写评价页面如图 9.2 和图 9.3 所示。

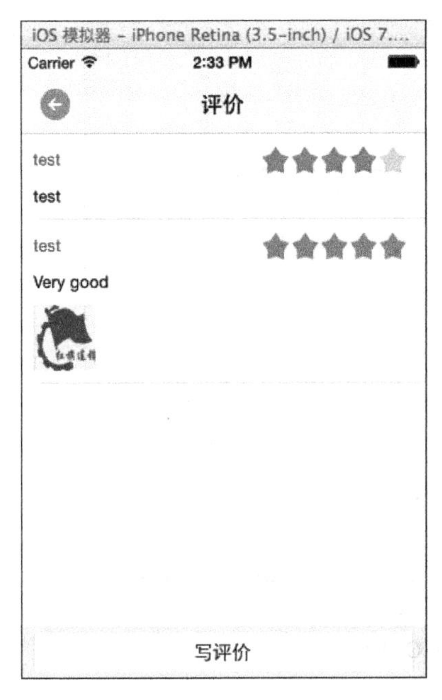

图 9.2 显示评价

图 9.3 写评价

9.3.1 客户端代码开发

1. 添加评价展示页面

(1) 选中 Sections 文件夹，按 command+N，选择 iOS 下的 Objective-C class，如图 9.4 所示。

(2) 在 Class 框中填写 AppraiseViewController，在 SubClass of 框中填写 FatherViewController，不要选中 Also create XIB file，如图 9.5 所示。

(3) 单击 AppraiseViewController.h，修改代码如下：

```
#import "FatherViewController.h"

@interface AppraiseViewController: FatherViewController

@property (nonatomic,copy)NSString * shopId;
@end
```

代码解析：

新建 shopId 属性，用来接收商户 id。后面会根据商户 id 来查询评价内容。

2. 添加显示评分视图

UITableViewCell 中需要用到显示评分视图。当用户点击或滑动评分视图时，五星

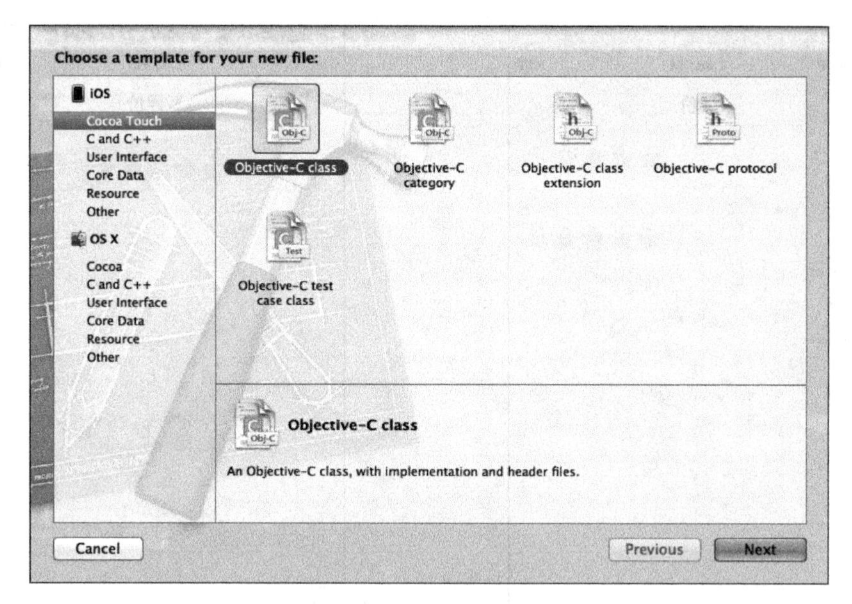

图 9.4 选择 Objective-C class

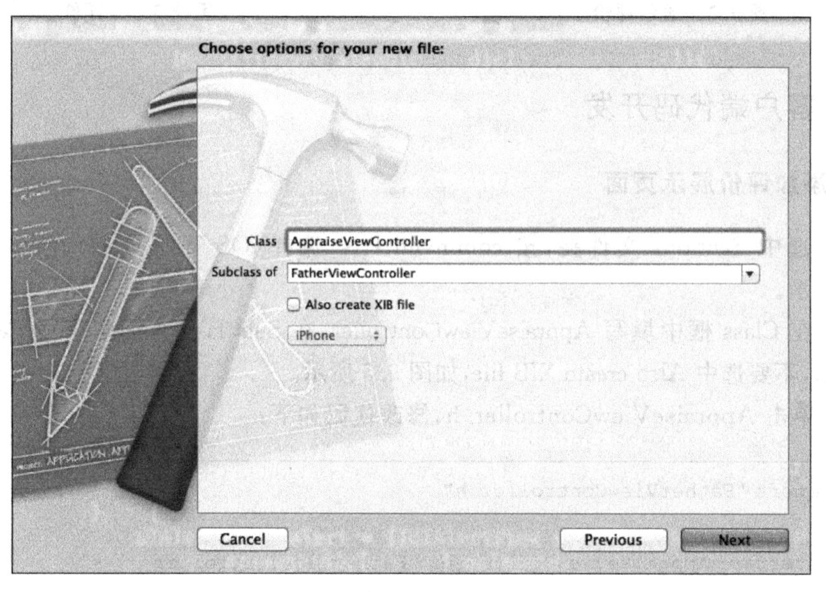

图 9.5 填写信息

高亮显示并实现打分,1 个高亮五星为 1 分。

要添加评分视图,可以在本书的电子资源中找到 RatingView. h 和 RatingView. m 两个文件并拖放到项目中,也可以手动创建。这里介绍如何手动创建。

(1) 右击 MyDemo 文件夹,选择 New Group,将新建的文件夹命名为 General,如图 9.6 所示。

(2) 选中 General 文件夹,按下 command+N,选择 iOS 下的 Objective-C class,如图 9.7 所示。

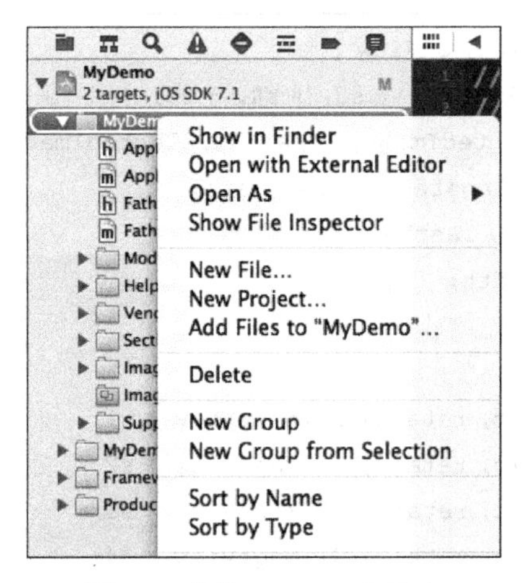

图 9.6 添加 General 文件夹

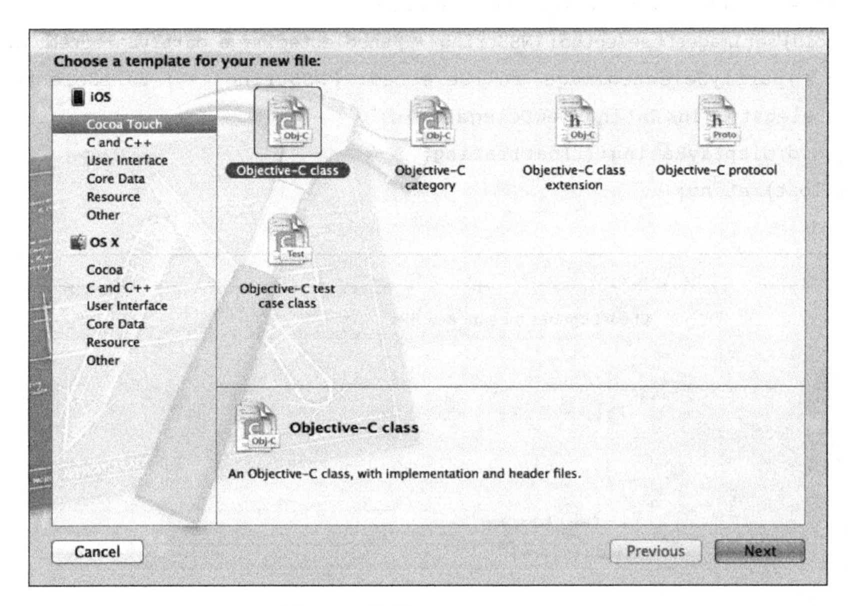

图 9.7 选择 Objective-C class

(3) 在 Class 中填写 RatingView,在 Subclass of 中填写 UIView,不要选中 Also create XIB file,如图 9.8 所示。

(4) 单击 RatingView.h,添加代码如下:

```
#import<UIKit/UIKit.h>

@protocol RatingViewDelegate
-(void)ratingChanged:(float)newRating;
@end
```

```
@interface RatingView: UIView {
    UIImageView * s1, * s2, * s3, * s4, * s5;
    UIImage * unselectedImage, * partlySelectedImage, * fullySelectedImage;
    id<RatingViewDelegate>viewDelegate;
    float starRating, lastRating;
    float height, width;
}

@property (nonatomic, retain) UIImageView * s1;
@property (nonatomic, retain) UIImageView * s2;
@property (nonatomic, retain) UIImageView * s3;
@property (nonatomic, retain) UIImageView * s4;
@property (nonatomic, retain) UIImageView * s5;
@property (nonatomic,assign)BOOL enable;
-(void)setImagesDeselected:(NSString *)unselectedImage partlySelected:(NSStr-
ing *)partlySelectedImage fullSelected:(NSString *) fullSelectedImage
andDelegate:(id<RatingViewDelegate>)d;
-(void)displayRating:(float)rating;
-(float)rating;
@end
```

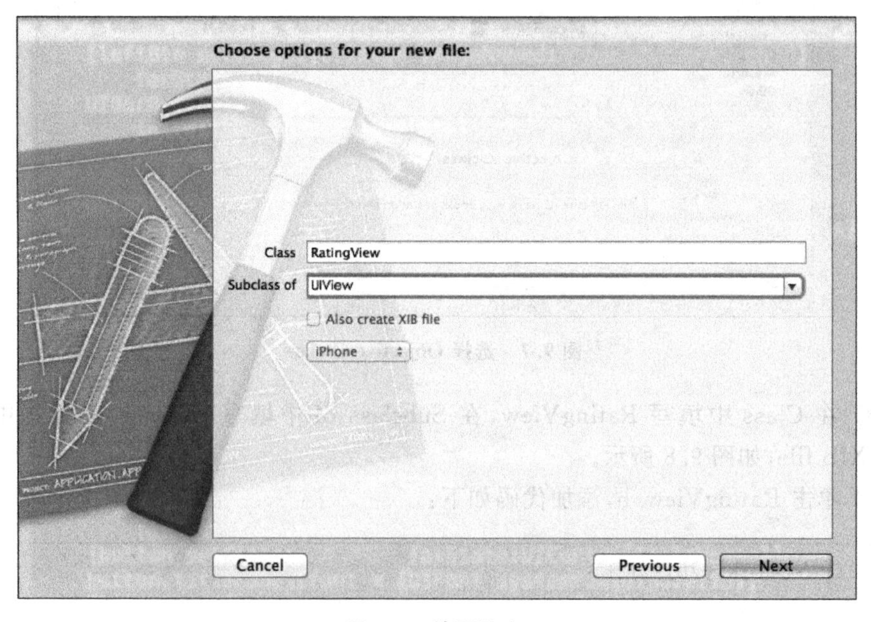

图 9.8　填写信息

代码解析：

enabl 属性用来开启或关闭评分视图的点击事件。因为页面有可能只需展示评分，不需要响应点击事件。

setImagesDeselected:partlySelected:fullSelected:andDelegate:方法用来设置不同状态的图片，分别是没有选中状态、半选中状态和选中状态。

displayRating:方法用来设置当前分数。

rating 方法返回当前分数。

(5) 单击 RatingView.m，添加代码如下：

```
#import "RatingView.h"
@implementation RatingView
@synthesize s1,s2,s3,s4,s5;
- (void) setImagesDeselected: (NSString *) deselectedImage partlySelected:
(NSString *)halfSelectedImage fullSelected:(NSString *)fullSelectedImage
andDelegate:(id<RatingViewDelegate>)d {
    if(!self.enable){
        self.userInteractionEnabled=NO;
    }
    unselectedImage=[UIImage imageNamed:deselectedImage];
    partlySelectedImage=halfSelectedImage==nil ? unselectedImage: [UIImage
imageNamed:halfSelectedImage];
    fullySelectedImage=[UIImage imageNamed:fullSelectedImage];
    viewDelegate=d;

    height=0.0; width=0.0;
    if (height<[fullySelectedImage size].height) {
        height=[fullySelectedImage size].height;
    }
    if (height<[partlySelectedImage size].height) {
        height=[partlySelectedImage size].height;
    }
    if (height<[unselectedImage size].height) {
        height=[unselectedImage size].height;
    }
    if (width<[fullySelectedImage size].width) {
        width=[fullySelectedImage size].width;
    }
    if (width<[partlySelectedImage size].width) {
        width=[partlySelectedImage size].width;
    }
    if (width<[unselectedImage size].width) {
        width=[unselectedImage size].width;
    }

    starRating=0;
```

```
    lastRating=0;
    s1=[[UIImageView alloc] initWithImage:unselectedImage];
    s2=[[UIImageView alloc] initWithImage:unselectedImage];
    s3=[[UIImageView alloc] initWithImage:unselectedImage];
    s4=[[UIImageView alloc] initWithImage:unselectedImage];
    s5=[[UIImageView alloc] initWithImage:unselectedImage];

    [s1 setFrame:CGRectMake(0,          0, width, height)];
    [s2 setFrame:CGRectMake(width,      0, width, height)];
    [s3 setFrame:CGRectMake(2 * width, 0, width, height)];
    [s4 setFrame:CGRectMake(3 * width, 0, width, height)];
    [s5 setFrame:CGRectMake(4 * width, 0, width, height)];

    [s1 setUserInteractionEnabled:NO];
    [s2 setUserInteractionEnabled:NO];
    [s3 setUserInteractionEnabled:NO];
    [s4 setUserInteractionEnabled:NO];
    [s5 setUserInteractionEnabled:NO];

    [self addSubview:s1];
    [self addSubview:s2];
    [self addSubview:s3];
    [self addSubview:s4];
    [self addSubview:s5];

    CGRect frame=[self frame];
    frame.size.width=width * 5;
    frame.size.height=height;
    [self setFrame:frame];
}

-(void)displayRating:(float)rating {
    [s1 setImage:unselectedImage];
    [s2 setImage:unselectedImage];
    [s3 setImage:unselectedImage];
    [s4 setImage:unselectedImage];
    [s5 setImage:unselectedImage];

    if (rating>=0.5) {
        [s1 setImage:partlySelectedImage];
```

```
    }
    if (rating>=1) {
        [s1 setImage:fullySelectedImage];
    }
    if (rating>=1.5) {
        [s2 setImage:partlySelectedImage];
    }
    if (rating>=2) {
        [s2 setImage:fullySelectedImage];
    }
    if (rating>=2.5) {
        [s3 setImage:partlySelectedImage];
    }
    if (rating>=3) {
        [s3 setImage:fullySelectedImage];
    }
    if (rating>=3.5) {
        [s4 setImage:partlySelectedImage];
    }
    if (rating>=4) {
        [s4 setImage:fullySelectedImage];
    }
    if (rating>=4.5) {
        [s5 setImage:partlySelectedImage];
    }
    if (rating>=5) {
        [s5 setImage:fullySelectedImage];
    }

    starRating=rating;
    lastRating=rating;
    [viewDelegate ratingChanged:rating];
}

- (void) touchesBegan: (NSSet *)touches withEvent: (UIEvent *)event
{
    [self touchesMoved:touches withEvent:event];
}
- (void) touchesMoved: (NSSet *)touches withEvent: (UIEvent *)event
{
    CGPoint pt=[[touches anyObject] locationInView:self];
```

```
    int newRating= (int) (pt.x / width)+1;
    if (newRating<1 || newRating>5)
        return;

    if (newRating !=lastRating)
        [self displayRating:newRating];
}

- (void)touchesEnded:(NSSet *)touches withEvent:(UIEvent *)event{
    [self touchesMoved:touches withEvent:event];
}

- (float)rating {
    return starRating;
}

@end
```

代码解析：

主要功能是当用户滑动或点击视图上的五星时，五星高亮显示并记录当前高亮的五星个数(1个高亮五星代表1分)。

3. 添加写评价页面

(1) 选中Sections文件夹，按下command+N，选择iOS下的Objective-C class，如图9.9所示。

(2) 在Class中填写WriteAppraiseViewController，在Subclass of中填写FatherViewController，不要选中Also create XIB file，如图9.10所示。

(3) 单击WriteAppraiseViewController.h文件，修改代码如下：

```
#import "FatherViewController.h"
typedef void(^CBBlock)(void);

@interface WriteAppraiseViewController: FatherViewController

@property (nonatomic,strong)NSString * shopId;
@property (nonatomic,strong)CBBlock block;

@end
```

代码解析：

这里创建一个block属性，用来实现评价后及时刷新页面数据。

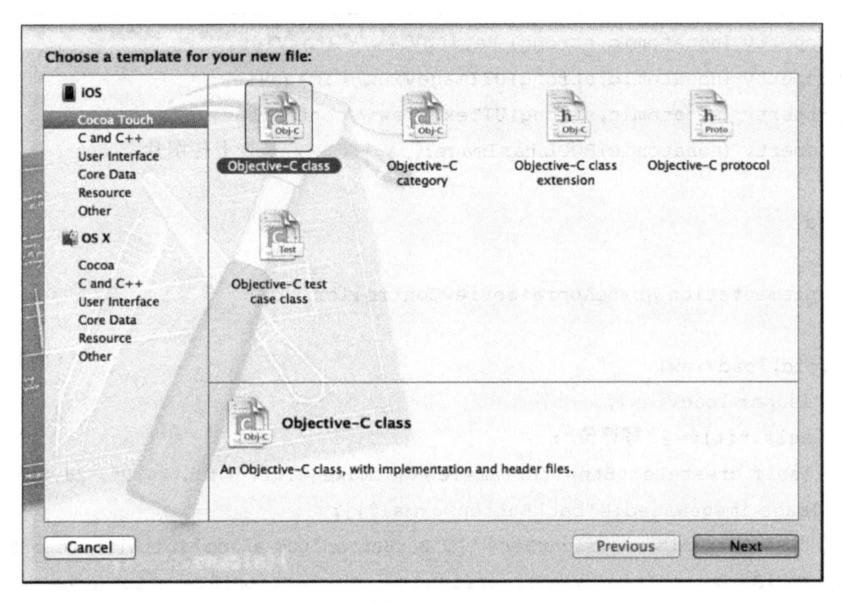

图 9.9 选择 Objective-C class

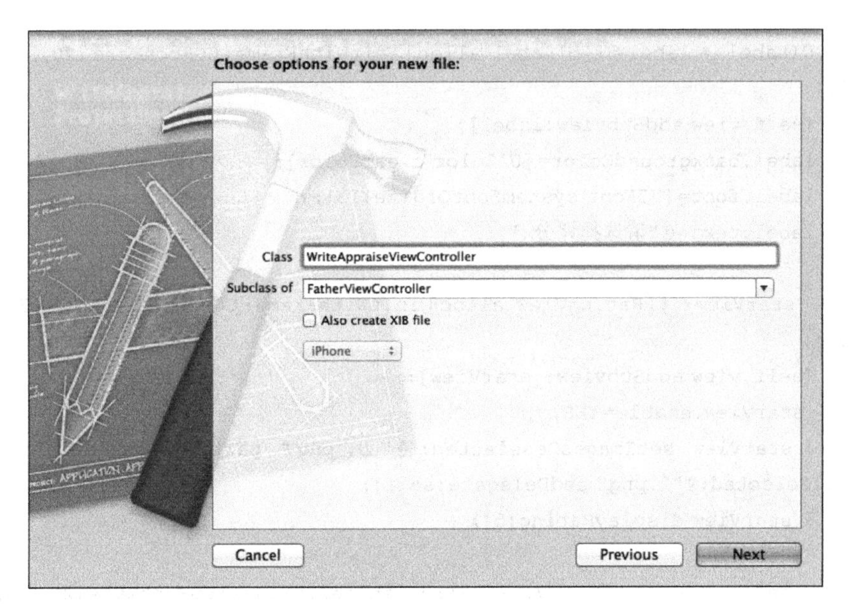

图 9.10 填写信息

(4) 单击 WriteAppraiseViewController.m 文件，修改代码如下：

```
#import "WriteAppraiseViewController.h"
#import "RatingView.h"
@interface WriteAppraiseViewController ()<RatingViewDelegate,UITextViewDelegate,
UIImagePickerControllerDelegate,UINavigationControllerDelegate>
```

```
@property (nonatomic,strong) RatingView * starView;
@property (nonatomic,strong)UIImageView * imageView;
@property (nonatomic,strong)UITextView * contentView;
@property (nonatomic)BOOL hasImage;          //是否上传图片

@end

@implementation WriteAppraiseViewController

- (void)loadView{
    [super loadView];
    self.title=@"写评价";
    [self createLeftBtnWithFrame:CGRectMake(0.0f, 0.0f, 24.0f, 24.0f) Image:
[UIImage imageNamed:@"backButtonNormal"]];
    UIBarButtonItem * rightBar= [[UIBarButtonItem alloc]initWithTitle:@"完成"
style:UIBarButtonItemStyleBordered target:self action:@selector(done)];
    self.navigationItem.rightBarButtonItem=rightBar;

    UILabel * label= [[UILabel alloc]initWithFrame:CGRectMake(20, 20, 100,
20)];
    [self.view addSubview:label];
    label.backgroundColor= [UIColor clearColor];
    label.font= [UIFont systemFontOfSize:13];
    label.text=@"请选择评分:";

    _starView= [[RatingView alloc] initWithFrame: CGRectMake (100, 20, 148,
65)];
    [self.view addSubview:_starView];
    _starView.enable=YES;
    [_starView setImagesDeselected: @"0.png" partlySelected: @"1.png"
fullSelected:@"2.png" andDelegate:self];
    [_starView displayRating:5];

    _contentView= [[UITextView alloc]initWithFrame:CGRectMake(20, 50, App_
SCREEN_WIDTH-40, 100)];
    [self.view addSubview:_contentView];
    _contentView.delegate=self;
    _contentView.textColor= [UIColor blackColor];
    _contentView.font= [UIFont systemFontOfSize:15];
    [_contentView becomeFirstResponder];
```

```
    _imageView= [[UIImageView alloc]initWithFrame:CGRectMake(20, 160, 100, 
100)];
    [self.view addSubview:_imageView];
    _imageView.image= [UIImage imageNamed:@"default"];

    UIButton * button= [UIButton buttonWithType:UIButtonTypeRoundedRect];
    [self.view addSubview:button];
    button.frame=CGRectMake(130, 230, 100, 30);
    [button setTitle:@"添加图片" forState:UIControlStateNormal];
    button.titleLabel.font= [UIFont systemFontOfSize:15];
    [button setTitleColor:[UIColor whiteColor] forState:UIControlStateNormal];
    [button setBackgroundImage: [UIImage imageNamed: @" checkopinion "] 
forState:UIControlStateNormal];
    [button addTarget:self action:@selector(selectImage) forControlEvents:
UIControlEventTouchUpInside];
}
- (void)leftBarChick{
    [self.navigationController popViewControllerAnimated:YES];
}
- (void)selectImage{
    UIImagePickerController * picker = [[UIImagePickerController alloc]
init];
    picker.delegate=self;
    picker.sourceType=UIImagePickerControllerSourceTypePhotoLibrary;
    picker.allowsEditing=YES;
    [self presentViewController:picker animated:YES completion:nil];
}
#pragma mark-UIImagePickerControllerDelegate
- (void) imagePickerController: (UIImagePickerController * ) picker 
didFinishPickingMediaWithInfo:(NSDictionary *)info
{
    UIImage * image= [info objectForKey:@"UIImagePickerControllerEditedImage"];
    _imageView.image=image;
    _hasImage=YES;
    [picker dismissViewControllerAnimated:YES completion:^{}];
}

- (void)imagePickerControllerDIdCancel:(UIImagePickerController*)picker
{
    [picker dismissViewControllerAnimated:YES completion:nil];
}
- (BOOL)textView:(UITextView *)textView shouldChangeTextInRange:(NSRange) 
range replacementText:(NSString *)text
```

```
{
    if ([text isEqualToString:@"\n"]) {
        [textView resignFirstResponder];
        return NO;
    }
    return YES;
}
- (void)ratingChanged:(float)newRating {
    NSLog(@"%f",_starView.rating);
}
- (void)done{
}
- (void)viewDidLoad
{
    [super viewDidLoad];
}
- (void)didReceiveMemoryWarning
{
    [super didReceiveMemoryWarning];
    //Dispose of any resources that can be recreated.
}
@end
```

代码解析：

hasImage属性用来记录用户是否选择了图片，默认为NO。

当用户单击选择图片按钮时，程序调用系统相册。用户选择完图片后，hasImage的值为YES。这样，当用户单击完成按钮上传评价信息时，程序就会先将图片上传到服务端，待服务端返回上传成功图片的地址后，程序再将图片地址和评价内容上传服务端。如果用户没有选择图片，那么hasImage为NO，程序直接上传评价内容到服务端。

UIImagePickerController控件提供用户调用相机或者本地相册的操作。当调用相机的时候，程序要验证相机设备是否可用，因为并不是所有苹果设备都有相机功能，因此有必要验证。本项目由于没有用到相机，因此不需要验证。

9.3.2 客户端与服务端交互

1. 添加评价Model类

（1）单击Models文件夹，按下command＋N，选择Cocoa Touch下的Objective-C class，如图9.11所示。

（2）在Class中填写AppraiseItem，在Subclass of中填写NSObject，如图9.12所示。

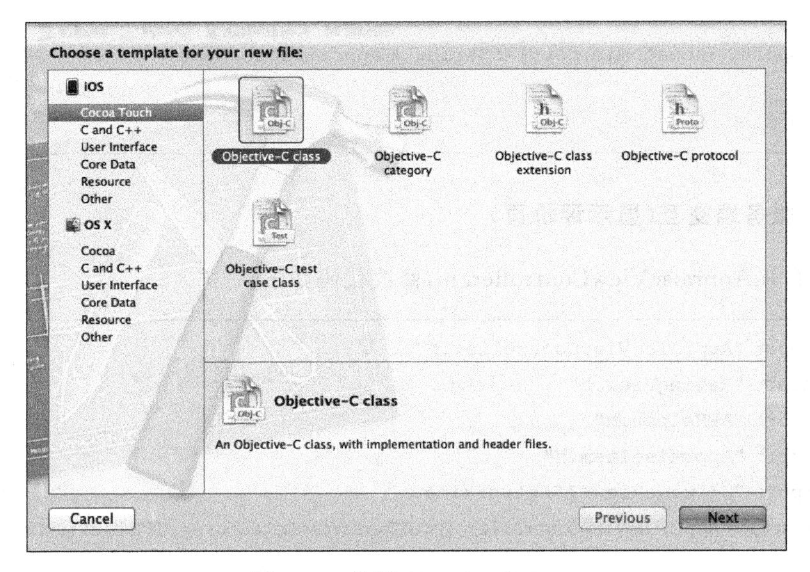

图 9.11　选择 Objective-C class

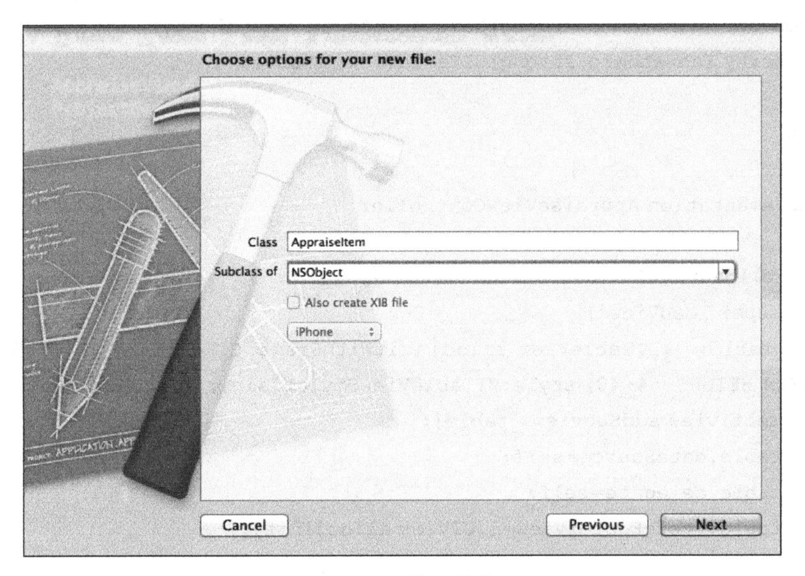

图 9.12　填写信息

(3) 单击 AppraiseItem.h,修改代码如下:

```
#import<Foundation/Foundation.h>

@interface AppraiseItem: NSObject
@property (nonatomic,copy)NSString *appraiseContent;
@property (nonatomic,copy)NSString *appraiseId;
@property (nonatomic,copy)NSString *appraiseImageUrl;
@property (nonatomic,copy)NSString *appraisePoint;
```

```
@property (nonatomic,copy)NSString * appraiseUserName;

@end
```

2. 与服务端交互(显示评价页)

(1) 单击 AppraiseViewController.m,修改代码如下:

```
#import "AppraiseViewController.h"
#import "RatingView.h"
#import "AFHelper.h"
#import "AppraiseItem.h"
#import "UIImageView+AFNetworking.h"
@interface AppraiseViewController ()<UITableViewDataSource,UITableViewDelegate>

@property (nonatomic, strong) RatingView * starView;
@property (nonatomic,strong)NSMutableArray * dataArray;
@property (nonatomic,strong)UITableView * table;

@end

@implementation AppraiseViewController

- (void)loadView{
    [super loadView];
    _table= [[UITableView alloc] initWithFrame:CGRectMake (0, 0, 320, App_
SCREEN_HEIGHT-64-40) style:UITableViewStylePlain];
    [self.view addSubview:_table];
    _table.dataSource=self;
    _table.delegate=self;
    _table.tableFooterView= [[UIView alloc]init];

    UIButton * button= [UIButton buttonWithType:UIButtonTypeCustom];
    button.frame=CGRectMake(10, App_SCREEN_HEIGHT-64-40+5, 300, 30);
    [self.view addSubview:button];
    [button setBackgroundImage: [UIImage imageNamed: @" findbg"] forState:
UIControlStateNormal];
    [button addTarget: self action: @ selector (writeAppraise) forContro-
lEvents:UIControlEventTouchUpInside];
    [button setTitle:@"写评价" forState:UIControlStateNormal];
    [button setTitleColor: [ UIColor blackColor ] forState: UIControl
StateNormal];
    button.titleLabel.font= [UIFont systemFontOfSize:15];
```

```
}
- (void)viewDidLoad
{
    [super viewDidLoad];
    self.title=@"评价";
    [self createLeftBtnWithFrame:CGRectMake(0.0f, 0.0f, 24.0f, 24.0f) Image:
[UIImage imageNamed:@"backButtonNormal"]];
    _dataArray= [NSMutableArray array];
    [self getData];
}
- (void)getData{
    NSDictionary * dic= @ {@"act":@"getAppraise",@"shopId":_shopId,@
"pageSize":@"20",@"page":@"1"};
    [AFHelper downDataWithDictionary:dic andBaseURLStr:@"http://localhost:
8080/meServer/" andPostPath:@"appraise.php?" success:^(NSDictionary * dic){
        if([dic[@"flag"]intValue]==1){
            for(NSDictionary * tmpDic in dic[@"commentList"]){
                AppraiseItem * appraiseItem= [[AppraiseItem alloc]init];
                appraiseItem.appraiseContent=tmpDic[@"appraiseContent"];
                appraiseItem.appraiseId=tmpDic[@"appraiseId"];
                appraiseItem.appraiseImageUrl=tmpDic[@"appraiseImageUrl"];
                appraiseItem.appraisePoint=tmpDic[@"appraisePoint"];
                appraiseItem.appraiseUserName=tmpDic[@"userName"];
                [_dataArray addObject:appraiseItem];
            }
            [_table reloadData];
        }else{
            UIAlertView * alert = [[UIAlertView alloc] initWithTitle: @"提示"
message:dic[@"msg"] delegate:nil cancelButtonTitle:@"好的" otherButtonTitles:
nil, nil];
            [alert show];
        }


    }];
}
- (NSInteger) tableView: (UITableView * ) tableView numberOfRowsInSection:
(NSInteger)section{
    return _dataArray.count;
}
- (CGFloat) tableView: (UITableView * ) tableView heightForRowAtIndexPath:
(NSIndexPath * )indexPath
```

```
{
    UITableViewCell * cell= [self tableView:tableView cellFor RowAtIndex
Path:indexPath];
    return cell.size.height;
}
- (UITableViewCell *) tableView: (UITableView *) tableView cellFor
RowAtIndexPath:(NSIndexPath *)indexPath{
    static NSString *CellIdentifier=@"appraiseCell";
    UITableViewCell * cell= [tableView dequeueReusableCellWithIdentifier:
CellIdentifier];
    if(cell==nil){
        cell=[[UITableViewCell alloc]initWithStyle:UITableViewCellStyleDefault
reuseIdentifier:CellIdentifier];
    }
    for(UIView *view in cell.contentView.subviews){
        [view removeFromSuperview];
    }
    AppraiseItem *appraiseItem=_dataArray[indexPath.row];
    UILabel *idLbl=[[UILabel alloc]initWithFrame:CGRectMake(10, 10, 100, 20)];
    [cell.contentView addSubview:idLbl];
    idLbl.textColor=[UIColor redColor];
    idLbl.font=[UIFont systemFontOfSize:13];
    idLbl.backgroundColor=[UIColor clearColor];
    idLbl.text=appraiseItem.appraiseUserName;

    _starView=[[RatingView alloc]initWithFrame:CGRectMake(190, 10, 50, 15)];
    [cell.contentView addSubview:_starView];
    [_starView setImagesDeselected:@"0" partlySelected:@"1" fullSelected:@
"2" andDelegate:nil];
    [_starView displayRating:[appraiseItem.appraisePoint floatValue]];

    UILabel * contentLbl = [[UILabel alloc] initWithFrame: CGRectMake (10,
idLbl.bottom+10, App_SCREEN_WIDTH-20, 20)];
    [cell.contentView addSubview:contentLbl];
    contentLbl.textColor=[UIColor blackColor];
    contentLbl.font=[UIFont systemFontOfSize:13];
    contentLbl.backgroundColor=[UIColor clearColor];
    contentLbl.numberOfLines=0;
    CGSize sizeToFit = [appraiseItem. appraiseContent sizeWithFont: [UIFont
systemFontOfSize: 13] constrainedToSize: CGSizeMake (320, 2000) lineBreakMode:
NSLineBreakByWordWrapping];
    contentLbl.frame=CGRectMake( contentLbl. left, contentLbl. top,
contentLbl.width,sizeToFit.height);
    contentLbl.text=appraiseItem.appraiseContent;
```

```
    CGRect rect=cell.frame;
    if(appraiseItem.appraiseImageUrl.length<1){
        rect.size.height=contentLbl.bottom+10;
    }else{
        UIImageView * imageview=[[UIImageView alloc]init];
        imageview.frame=CGRectMake(contentLbl.left, contentLbl.bottom+10,
50, 50);
        [cell.contentView addSubview:imageview];
        [imageview setImageWithURL: [NSURL URLWithString: appraiseItem.
appraiseImageUrl]];
        rect.size.height=imageview.bottom+10;
    }
    cell.frame=rect;
    return cell;
}
- (void) tableView: (UITableView * ) tableView didSelectRowAtIndexPath:
(NSIndexPath * )indexPath{
}
-(void)leftBarChick{
    [self.navigationController popViewControllerAnimated:YES];
}
-(void)didReceiveMemoryWarning
{
    [super didReceiveMemoryWarning];
    //Dispose of any resources that can be recreated.
}
-(void)writeAppraise{
}
@end
```

代码解析：

要注意 tableView:cellForRowAtIndexPath:方法。

这里用纯代码来创建自定义 Cell。由于评价内容有长有短、图片可有可无，所以 Cell 的高度不能像以前一样固定，而是要根据评价内容动态调整高度。评价内容用一个 UILabel 控件显示并根据文本内容动态设置 UILabel 的高度。代码如下：

```
UILabel * contentLbl = [[UILabel alloc]initWithFrame:CGRectMake(10, idLbl.
bottom+10, App_SCREEN_WIDTH-20, 20)];
    [cell.contentView addSubview:contentLbl];
    contentLbl.textColor=[UIColor blackColor];
    contentLbl.font=[UIFont systemFontOfSize:13];
    contentLbl.backgroundColor=[UIColor clearColor];
```

```
    contentLbl.numberOfLines=0;
    CGSize sizeToFit = [appraiseItem.appraiseContent sizeWithFont: [UIFont
systemFontOfSize: 13] constrainedToSize: CGSizeMake(320, 2000) lineBreakMode:
NSLineBreakByWordWrapping];
    contentLbl.frame=CGRectMake(contentLbl.left, contentLbl.top, contentLbl.
width,sizeToFit.height);
    contentLbl.text=appraiseItem.appraiseContent;
```

获取了Cell内容的高度后，需要设置Cell的高度。如果评价内容没有图片，那么Cell的高度就是UILabel的高度＋UILabel相对于Cell的y坐标。如果有图片，则Cell的高度就是UILabel的高度＋UILabel相对于Cell的y坐标＋图片的高度，代码如下：

```
    CGRect rect=cell.frame;
    if(appraiseItem.appraiseImageUrl.length<1){
        rect.size.height=contentLbl.bottom+10;
    }else{
        UIImageView * imageview=[[UIImageView alloc]init];
        imageview.frame=CGRectMake(contentLbl.left, contentLbl.bottom+10,
50, 50);
        [cell.contentView addSubview:imageview];
        [imageview setImageWithURL: [NSURL URLWithString: appraiseItem.
appraiseImageUrl]];
        rect.size.height=imageview.bottom+10;
    }
    cell.frame=rect;
```

最后在tableView: heightForRowAtIndexPath方法中返回Cell的高度，代码如下：

```
- (CGFloat) tableView: (UITableView * ) tableView heightForRowAtIndexPath:
(NSIndexPath * )indexPath
{
    UITableViewCell * cell = [self tableView: tableView cellForRowAtIndexPath:
indexPath];
    return cell.size.height;
}
```

(2) 添加显示评价页面入口。单击ShopDetailViewController.m文件，引入显示评价页头文件并重写goEvaluateVC:方法。代码如下：

```
#import "AppraiseViewController.h"

-(IBAction)goEvaluateVC:(id)sender {
    AppraiseViewController * vc=[[AppraiseViewController alloc]init];
    vc.shopId=_shopItem.shopId;
    [self.navigationController pushViewController:vc animated:YES];
}
```

3. 与服务端交互(写评价页面)

(1) 单击 WriteAppraiseViewController.m,引入头文件,代码如下:

```
#import "AFHelper.h"
#import "AFNetworking.h"
```

(2) 重写 done 方法,代码如下:

```
-(void)done{
    if(_contentView.text.length<1){
        UIAlertView * alert=[[UIAlertView alloc]initWithTitle:@"提示" message:
@"请输入评价内容" delegate:nil cancelButtonTitle:@"好的" otherButtonTitles:nil,
nil];
        [alert show];
    }else{
        if(_hasImage){
            NSData * imageData=UIImageJPEGRepresentation(_imageView.image,
0.5);
            NSDictionary * dic=@{@"act":@"postImage"};
            [AFHelper postImageWithDictionary: dic andImageData: imageData
andImageName: @" 1 " andBaseURLStr: @" http://localhost: 8080/meServer/"
andPostPath:@"image.php?" success:^(NSDictionary * dic){
                NSDictionary * dic1 = @ {@" act ": @" postAppraise ", @
"appraiseContent ": _ contentView. text, @" appraisePoint ": [ NSString
stringWithFormat:@"%f",[_starView rating]],@"appraiseImageUrl":[NSString
stringWithFormat: @"% @% @", @" http://localhost: 8080/meServer ", dic [@
"image"]],@"shopId":_shopId,@"userName":@"test"};
                [AFHelper downDataWithDictionary: dic1 andBaseURLStr: @" http:
//localhost:8080/meServer/" andPostPath:@"appraise.php?" success:^(NSDiction-
ary * dic){
```

```
                    self.block();
                    UIAlertView *alert=[[UIAlertView alloc]initWithTitle:@"提示"
message:dic[@"msg"] delegate:nil cancelButtonTitle:@"好的" otherButtonTitles:nil,
nil];
                    [alert show];
                }];
            }];
        }else{
            NSDictionary *dic=@{@"act":@"postAppraise",@"appraiseContent":
_contentView.text,@"appraisePoint":[NSString stringWithFormat:@"%f",[_starView
rating]],@"appraiseImageUrl":@"",@"shopId":_shopId,@"userName":@"test"};
            [AFHelper downDataWithDictionary: dic andBaseURLStr: @" http:
//localhost:8080/meServer/" andPostPath:@"appraise.php?" success:^(NSDiction-
ary *dic){
                self.block();
                UIAlertView *alert=[[UIAlertView alloc]initWithTitle:@"提示"
message:dic[@"msg"] delegate:nil cancelButtonTitle:@"好的" otherButtonTitles:
nil, nil];
                [alert show];
            }];
        }
    }
}
```

代码解析：

由于用户可能要上传图片，因此需要两个接口。如果用户选择了图片，则程序先上传图片到服务端，上传成功后，记录服务端返回的图片地址并再次上传评价内容（评分、评价文字、图片地址）到服务端。如果用户没有选择图片，那么直接上传评价内容（评分、评价文字）到服务端。

当用户提交评论内容成功后，程序调用 block()，因此可以在入口函数编写 block 的实现。

(3) 添加写评论页面入口。单击 AppraiseViewController.m 文件，引入写评价页头文件和登录页头文件，重写 writeAppraise 方法，代码如下：

```
#import "WriteAppraiseViewController.h"
#import "LoginViewController.h"

-(void)writeAppraise{
    if([USER_DEFAULT boolForKey:@"isLogin"]){
        WriteAppraiseViewController * vc = [[WriteAppraiseViewController
alloc]init];
        vc.block=^{
```

```
            [self getData];
        };
        vc.shopId=_shopId;
        [self.navigationController pushViewController:vc animated:YES];
    }else{
        LoginViewController * vc= [[LoginViewController alloc] initWithNibName:
@"LoginViewController" bundle:nil];
        vc.title=@"登录";
        UINavigationController * nav= [[UINavigationController alloc] init-
WithRootViewController:vc];
        [self presentViewController:nav animated:YES completion:nil];
    }
}
```

代码解析：

这里程序做了一个判断，当用户单击写评论按钮时，如果用户登录，则跳转到写评论页面；如果没有登录，则跳转到登录页面。

当用户提交评论成功后，显示评价页面就会响应 block，刷新评价内容。

9.4 让用户分享

本节介绍 App 中非常重要的一个功能：分享。完成后的界面如图 9.13 所示。

9.4.1 什么是分享

在智能手机应用中已有分享功能。比如听到一首歌想分享给好友，就可以通过应用中的分享功能，分享到新浪微博、腾讯微博、QQ 空间、微信等社交平台，呼叫好友围观、单击、转发、评价等。

这样就产生了一个社交传播闭环，更多用户看到分享的内容，分享评价的时候，会覆盖越来越多的粉丝，覆盖的粉丝会产生非常多的回流单击，回流单击又会产生社交单击，社交单击又会产生更多分享。

图 9.13 分享界面

9.4.2 让用户将内容分享到社交平台

开发者通常会自己开发这个分享功能，但由于每个分享平台需要分别配置，往往需要大量烦琐的工作。如果利用 App 分享功能组件，直接粘贴一段代码就可以集成，从而为开发者节省了许多精力和时间。

下面介绍 iOS 平台上面非常热门的一款分享组件：ShareSDK。

ShareSDK 是社会化分享组件，为 iOS、Android、WP8 的 App 提供了社会化分享功能，集成了一些常用的类库和接口，缩短了开发者的开发时间，并且还有社会化统计分析管理后台，支持包括 QQ、微信、新浪微博、腾讯微博、开心网、人人网、豆瓣、网易微博、搜狐微博、facebook、twitter、google+等国内外 40 多家的主流社交平台，可帮助开发者轻松实现社会化分享、登录、关注以及获得用户资料和获取好友列表等主流的社会化功能。强大的统计分析管理后台可以实时了解用户、信息流、回流率、传播效率等数据，有效指导移动 App 的日常运营与推广，同时为 App 引入更多的社会化流量。

集成 ShareSDK 的步骤如下：

(1) 访问 ShareSDK 官网 http://sharesdk.mob.com/Download 下载 SDK，如图 9.14 所示。

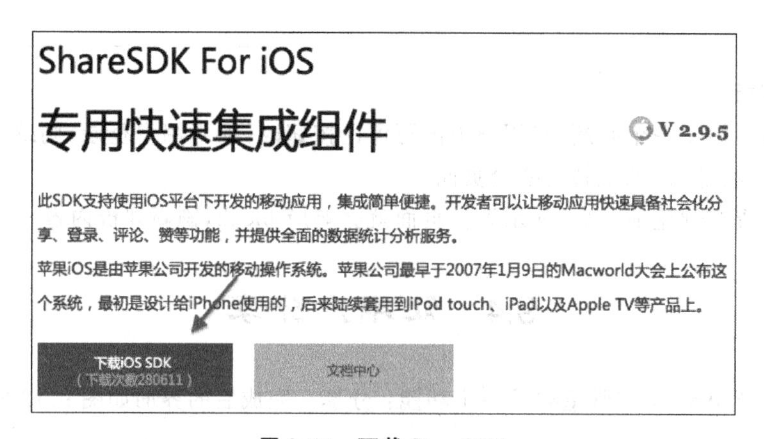

图 9.14　下载 ShareSDK

(2) 将下载后的文件解压，直接拖放解压后的 ShareSDK 文件夹到项目 Venders 文件夹中，注意选中 Copy items into destination group's folder(if needed)，如图 9.15 所示。

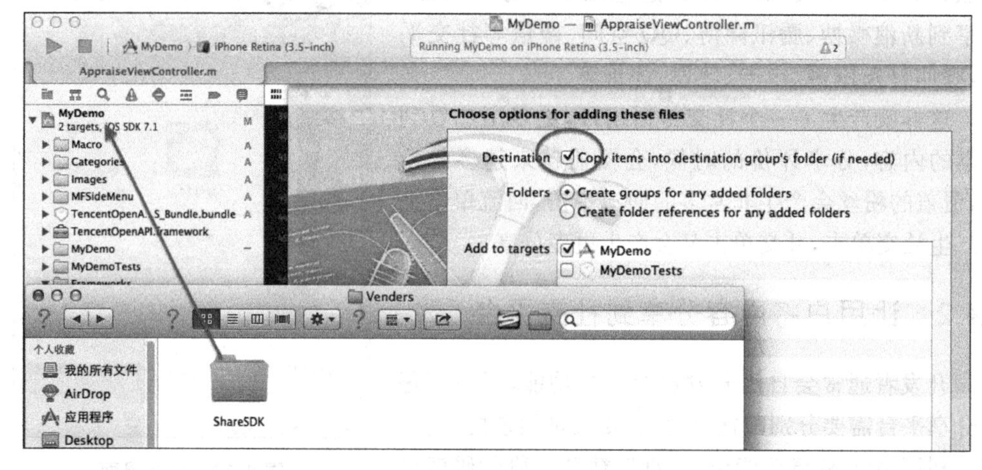

图 9.15　拖放 ShareSDK 到项目中

(3) 添加下列依赖框架,如图 9.16 所示。

```
QuartzCore.framework
libicucore.dylib
libz.1.2.5.dylib。
```

图 9.16 添加依赖框架

(4) 引入头文件和注册 ShareSDK 的 AppKey。

项目中要用到 ShareSDK 分享功能,需要到 ShareSDK 官网注册并创建应用,每一个应用都有唯一的 AppKey。拿到 AppKey 后,就可以使用 ShareSDK 了。如果要用到微博、微信等分享平台,还需要到各分享平台注册账号并创建应用拿到唯一标示。这里为方便讲解,就用 ShareSDK Demo 中提供的唯一标示。

单击 AppleDate. h,导入文件头 ShareSDK. h 和微信头文件,代码如下:

```
#import<ShareSDK/ShareSDK.h>
#import "WXApi.h"
```

在 application: didFinishLaunchingWithOptions:方法中调用 registerApp 方法来初始化 SDK,代码如下:

```
- (BOOL) application: (UIApplication * ) application didFinishLaunching
WithOptions:(NSDictionary * )launchOptions
{
    [ShareSDK registerApp:@"api20"];
                                    //参数为 ShareSDK 官网中添加应用后得到的 AppKey
    [self initializePlat];
    [self initializePlatForTrusteeship];

    //其他代码……
    return YES;
}
```

```
-(void)initializePlat{
    //添加新浪微博应用
    [ShareSDK connectSinaWeiboWithAppKey: @" 568898243 " appSecret: @
"38a4f8204cc784f81f9f0daaf31e02e3" redirectUri:@"http://www.sharesdk.cn"];
    //添加微信应用
    [ShareSDK connectWeChatWithAppId: @" wx4868b35061f87885 " wechatCls:
[WXApi class]];
}

-(void)initializePlatForTrusteeship{
    //导入微信需要的外部库类型
    [ShareSDK importWeChatClass:[WXApi class]];
}
```

(5) 分享内容。

单击 ShopDetailViewController.m,导入头文件,代码如下:

```
#import<ShareSDK/ShareSDK.h>
```

创建右上角的分享按钮,代码如下:

```
-(void)viewDidLoad
{
    [super viewDidLoad];
    //其他代码
    [self createRightBtnWithFrame: CGRectMake (0.0f, 0.0f, 24.0f, 24.0f)
Image:[UIImage imageNamed:@"rightBarChick"]];
}
```

实现 rightBarChick 方法,代码如下:

```
-(void)rightBarChick{
    //分享的内容
    id<ISSContent > publishContent = [ ShareSDK content: @" 分 享 内 容 "
defaultContent:@"默认分享内容,没内容时显示" image:nil title:@"ShareSDK" url:
@"http://www.sharesdk.cn" description: @"这 是 一 条 测 试 信 息" mediaType:
SSPublishContentMediaTypeNews];
    //弹出分享菜单
    [ShareSDK showShareActionSheet: nil shareList: nil content: publishContent
statusBarTips:YES authOptions: nil shareOptions: nil result: ^ (ShareType type,
SSResponseState state, id< ISSPlatformShareInfo> statusInfo, id< ICMErrorInfo>
error, BOOL end) {
```

```
    if (state==SSResponseStateSuccess) {
        NSLog(@"分享成功");
    }
    else if (state==SSResponseStateFail) {
        NSLog(NSLocalizedString(@"TEXT_SHARE_FAI", @"发布失败!error code
==%d, error code==%@"), [error errorCode], [error errorDescription]);
      }
  }];

}
```

代码解析：

这里主要调用 ShareSDK 封装好的方法，开发者只需传送对应的参数即可。此外，ShareSDK 的分享样式支持自定义。如果需要，可以查阅 ShareSDK 的技术文档。

(6) 单击 Xcode 左上角的▶按钮运行程序。输入新浪微博账号分享后，提示新浪微博分享成功，如图 9.17 所示。

图 9.17 分享界面

9.5 给用户推送消息

通过苹果推送服务，开发者能主动、及时地向用户发起交互，向其发送聊天消息、日程提醒、活动预告、进度提示、动态更新等。精准的目标用户和有价值的推送内容可以提升用户忠诚度,提高留存率与收入。

9.5.1 推送原理

在实现消息推送之前先提及几个关于推送的相关概念。推送流程图如图 9.18 所示。

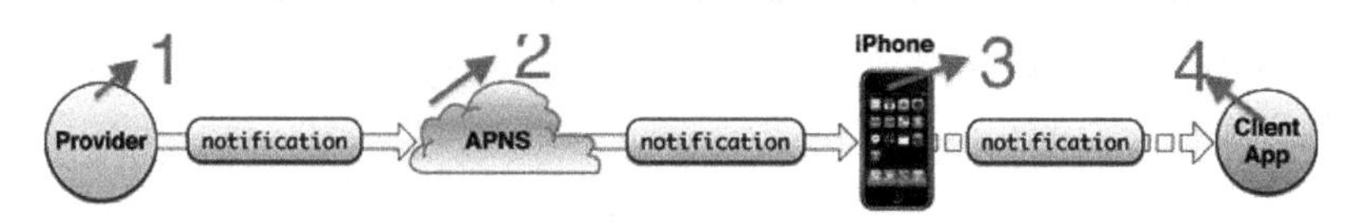

图 9.18 推送流程图

Provider: 为 iOS 设备应用程序提供推送的服务端(如果 iOS 设备的应用程序是客户端,那么 Provider 可以理解为服务端)。

APNS: 苹果推送通知服务(Apple Push Notification Service)。

iPhone: 用来接收 APNS 发送的消息。

Client App: iOS 设备上的应用程序,用来接收 iPhone 传递 APNS 下发的消息到制定的一个客户端 App。

图 9.18 可以分为三个阶段:

(1) Provider 把要发送的消息、目的 iOS 设备标识打包,发送给 APNS。

(2) APNS 在自身的已注册 Push 服务的 iOS 设备列表中,查找有相应标识的 iOS 设备,并将消息发送到 iOS 设备。

(3) iOS 设备把发送的消息传递给对应的应用程序,并按照设定弹出推送通知。

9.5.2 第三方推送介绍

目前,iOS 平台主流的第三方推送服务有极光推送、百度推送、个推、华为推送等。使用这些第三方推送,可以很大程度上简化 App 集成推送功能的复杂度,并方便在后台对推送信息进行设置。本节以极光推送和百度云推送为例,介绍二者的实现与区别。

1. 极光推送(JPush)

JPush 在 iOS 平台上有完整的推送服务,它整个推送过程完全不依赖 APNS 的服务,也就是把 APNS 变成了 JPush 自己的 Push 服务端。iPhone 到 Client App 的过程被简化了,JPush 采用的是透传方式,消息的传递对用户是透明的、不可见的,消息从 JPush 服务端直接就传到了 Client App,用户无法感知。

2. 百度云推送

百度云推送基于 APNS,也就是说它仅仅是 APNS 的一个代理,其推送流程如图 9.19 所示。

整个过程分为以下几个阶段:

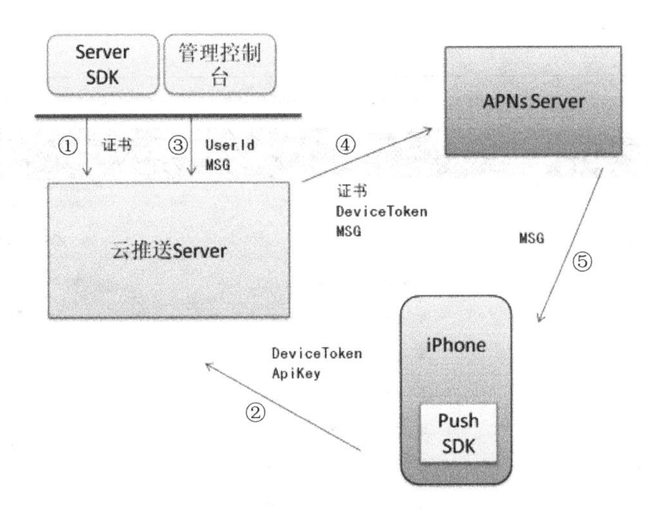

图 9.19 百度云推送流程

(1) 管理控制台或者 Server SDK 初始化 iOS App 的证书(分为开发版证书和发行版证书)。

(2) 运行在手机上的 Push SDK 执行推送的初始化动作,将 AppKey 和 DevicesToken 上传给云推送服务端,服务端保留。

(3) 管理控制台或者 Server SDK 向云推送服务端发送一条推送指令,服务端接到指令后,将控制台传来的 UserId(如果是广播,则没有 UserId)、Msg 与服务端保留的 DevicesToken 和证书一起打包传给 APNS 服务端。

(4) APNS 接到数据后,根据 UserId,将消息推送给指定的 iPhone 设备。

9.5.3 集成第三方推送

下面以极光推送为例,介绍在项目中如何集成极光推送。

1. 创建推送证书

(1) 打开 https://developer.apple.com/devcenter/ios/index.action,选择进入 iOS Provisioning Portal,如图 9.20 所示。

(2) 在 iOS Provisioning Portal 中,单击 App IDs 进入 App ID 列表,如图 9.21 所示。

(3) 创建 App ID。如果 ID 已经存在,则可以直接跳过此步骤,如图 9.22 所示。

(4) 为 App 开启 Push Notification 功能。如果是已经创建的 App ID,也可以通过设置开启 Push Notification 功能,如图 9.23 所示。

(5) 根据实际情况完善 App ID 信息并提交,注意此处需要指定具体的 Bundle ID,如图 9.24 所示。

(6) 如果之前没有创建过 Push 证书或者要重新创建新的证书,请在证书列表下面创建,如图 9.25 所示。

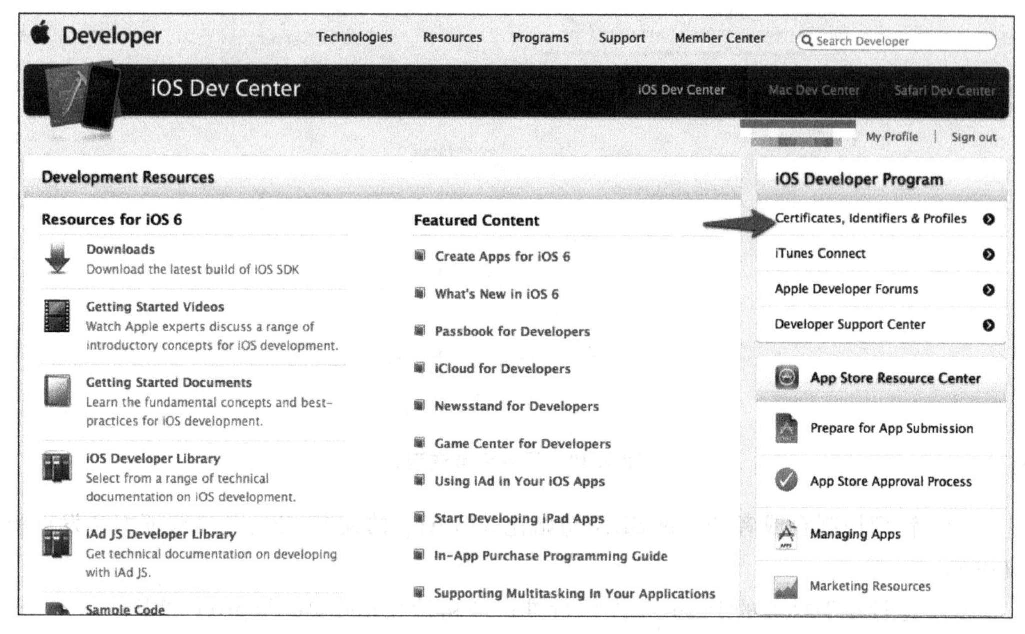

图 9.20　进入 iOS Provisioning Portal

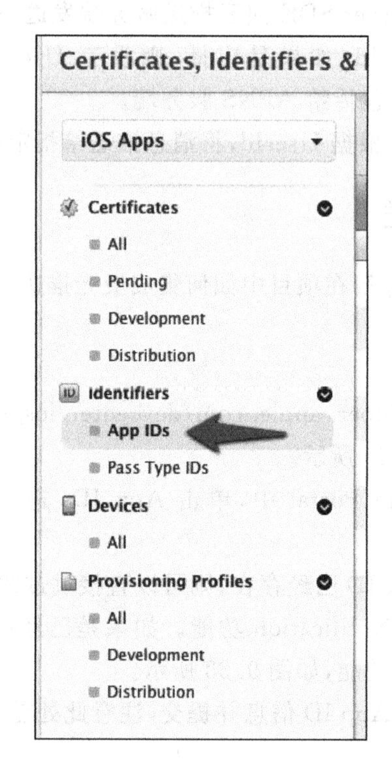

图 9.21　单击 App IDs

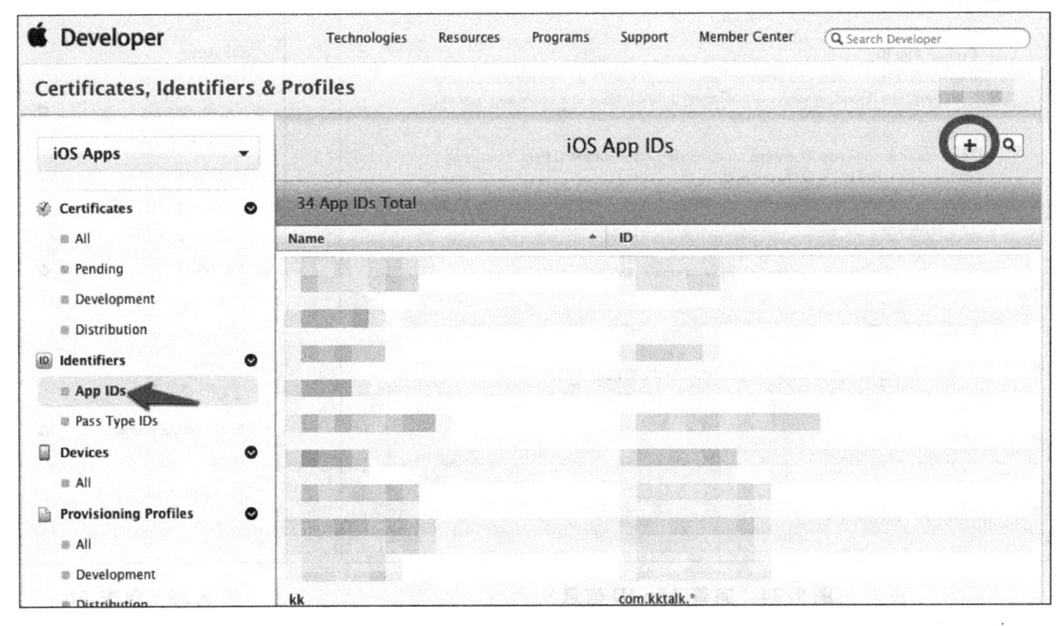

图 9.22 创建 App ID

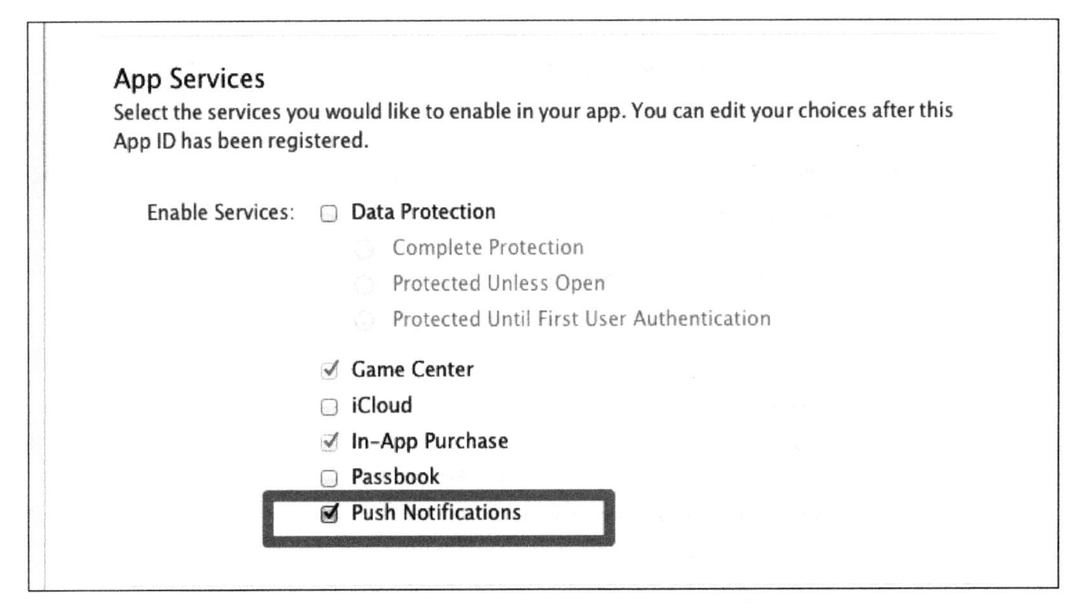

图 9.23 开启 Push Notification

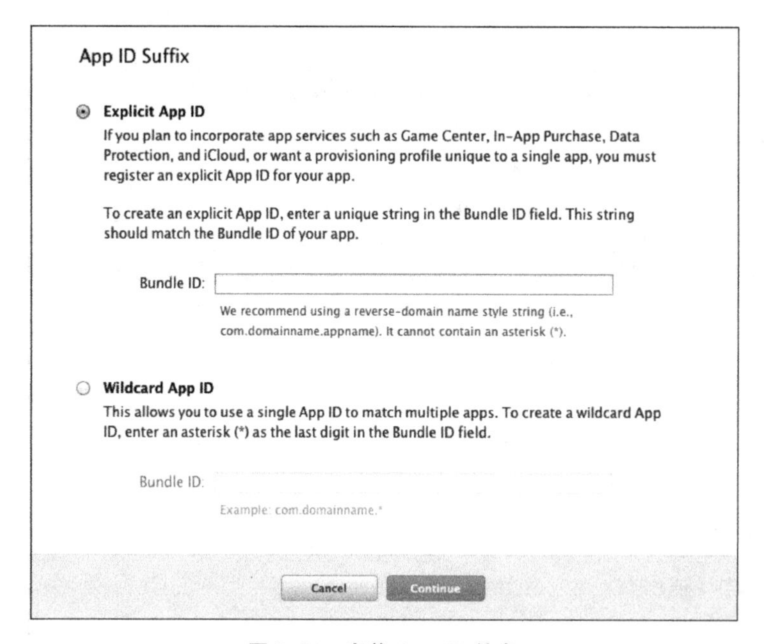

图 9.24　完善 App ID 信息

图 9.25　单击 All

(7) 新建证书需要注意选择证书种类，开发证书用于开发和调试使用，生产证书用于 App Store 发布，如图 9.26 所示。

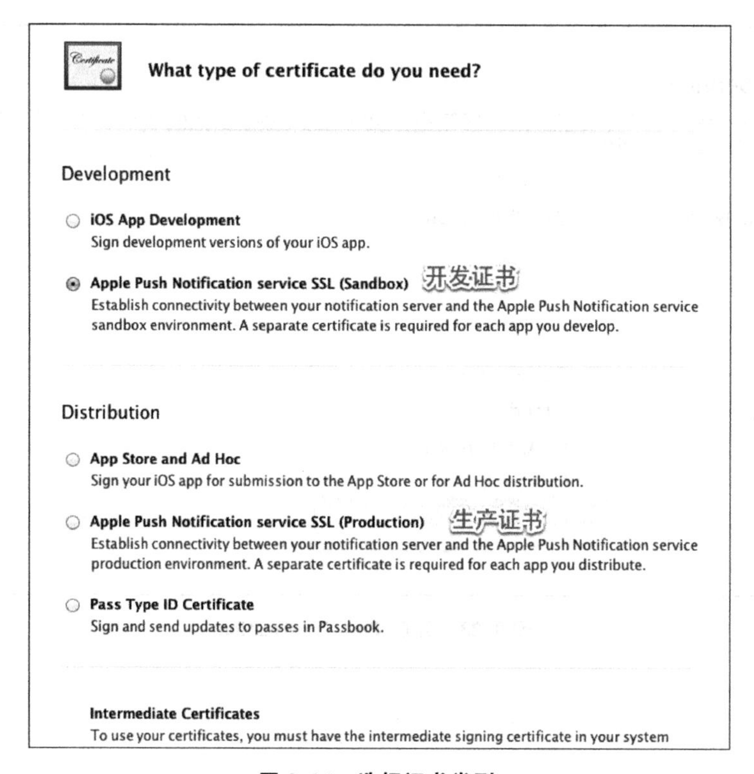

图 9.26　选择证书类型

(8) 单击 Continue 后选择证书对应的应用 ID，然后会继续出现 About Creating a Certificate Signing Request(CSR)，如图 9.27 所示。

图 9.27 创建 CSR

(9) 根据说明完成创建。打开 Keychain Access，创建 Certificate Signing Request，如图 9.28 所示。

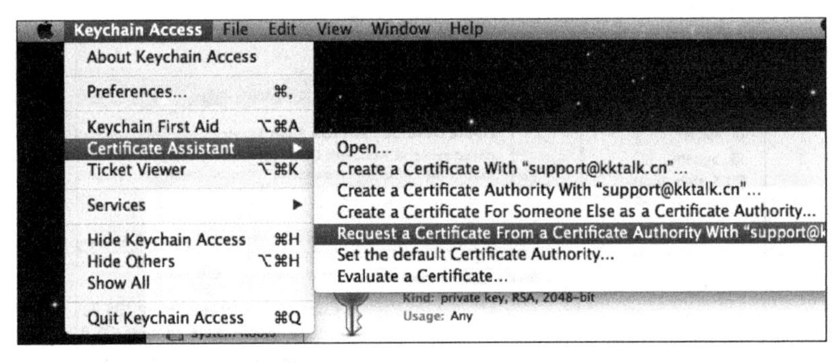

图 9.28 打开 Keychain Access

(10) 填写 User Email Address 和 Common Name，选择 Saved to disk 进行保存，如图 9.29 所示。

(11) 返回 Apple developer 网站，单击 Continue，上传刚刚生成的.certSigningRequest 文件并生成 APNs Push Certificate。下载后双击打开证书，证书打开时会启动“钥匙串访问”工具。在“钥匙串访问”中创建的证书会显示在“我的证书”中，注意选择 My Certificates 和 login，如图 9.30 所示。

(12) 在“钥匙串访问”中，选择刚刚加进来的证书。选择右键菜单中的 Export “Apple Development IOS Push Services”，如图 9.31 所示。

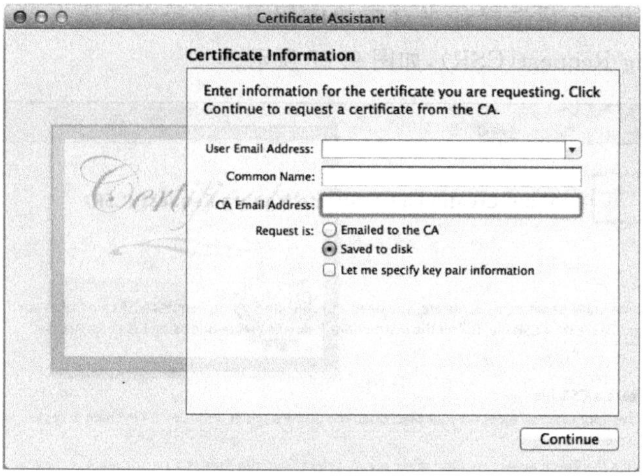

Certificate Assistant

Certificate Information

Enter information for the certificate you are requesting. Click Continue to request a certificate from the CA.

User Email Address:

Common Name:

CA Email Address:

Request is: Emailed to the CA

Saved to disk

Let me specify key pair information

Continue

图 9.29 填写信息

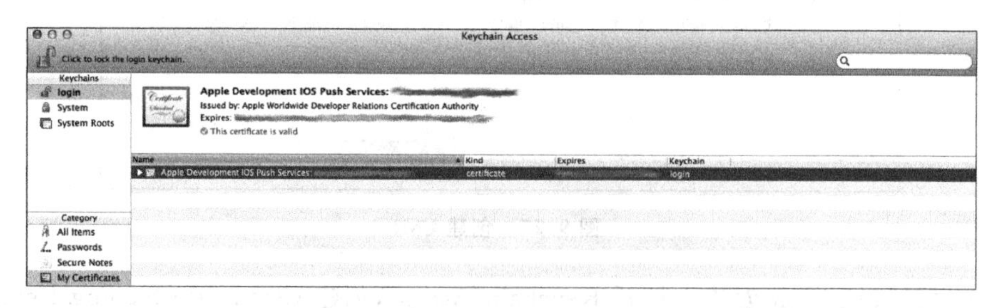

图 9.30 注意选择 My Certificates 和 login

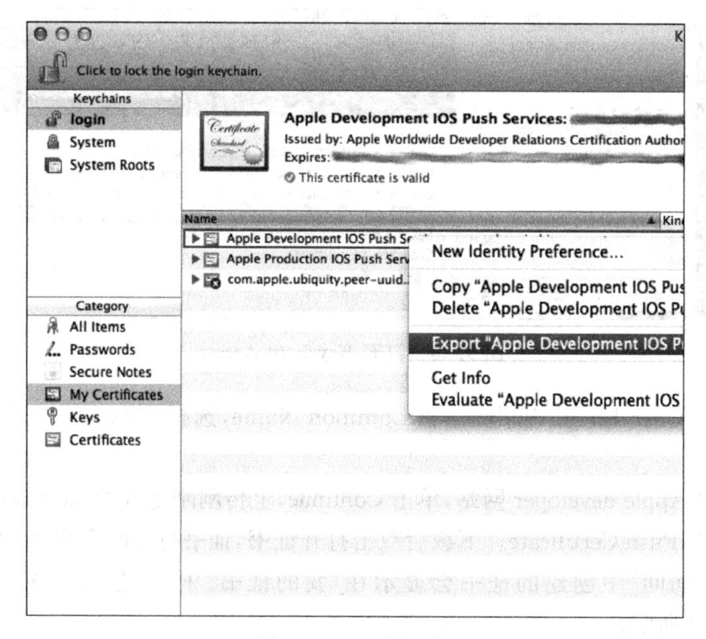

图 9.31 导出证书

(13) 将文件保存为 Personal Information Exchange(.p12)格式,如图 9.32 所示。

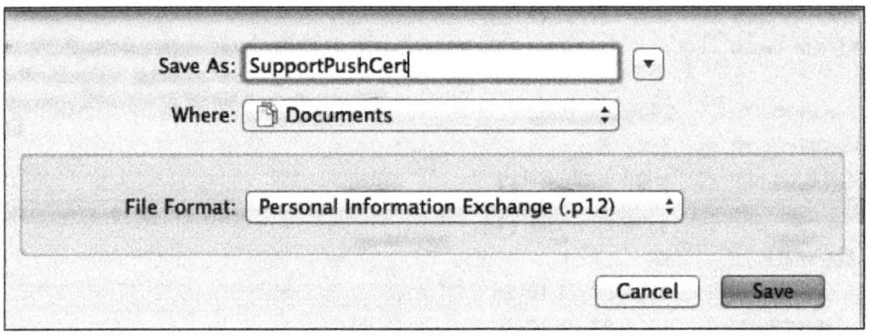

图 9.32 保存导出文件

2. 集成极光推送 SDK

(1) 打开 https://www.jpush.cn/common/apps,上传证书并创建应用,如图 9.33 所示。

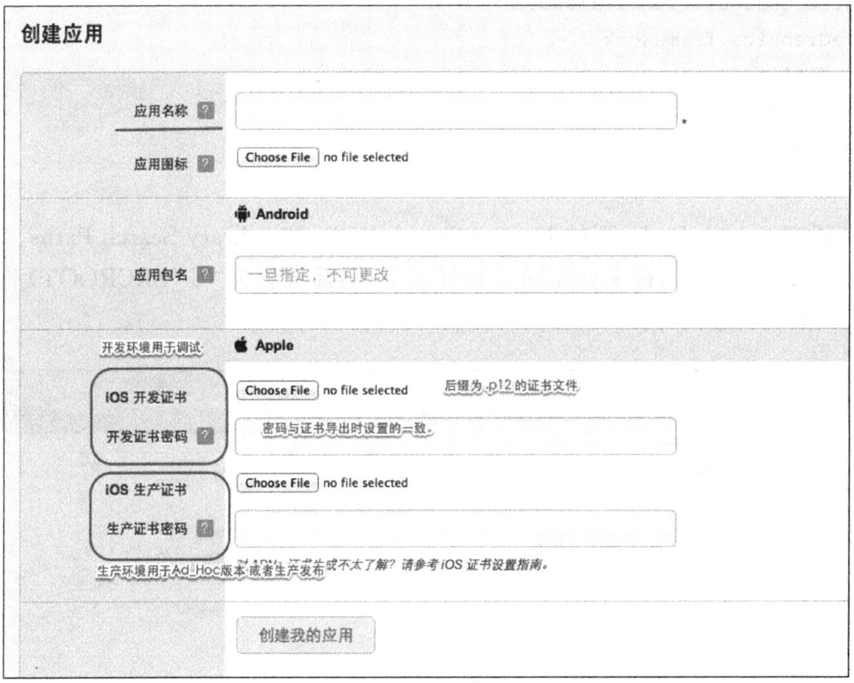

图 9.33 上传证书

(2) 创建成功后,自动生成 AppKey 用以标识该应用,如图 9.34 所示。

(3) 打开 http://docs.jpush.cn/display/dev/iOS,下载极光推送 iOS SDK,将解压后的 lib 子文件夹(包含 APService.h、libPushSDK.a)添加到工程目录中。

(4) 添加下列依赖框架:

```
CFNetwork.framework
CoreFoundation.framework
```

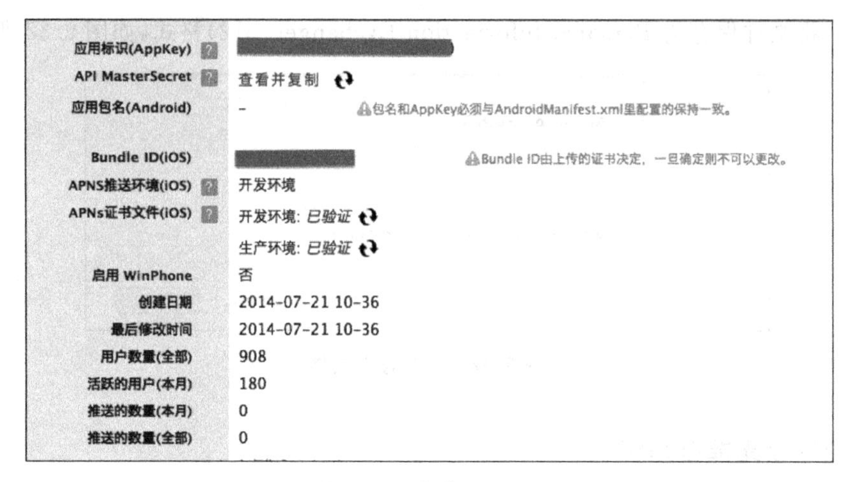

图 9.34　生成 AppKey

```
CoreTelephony.framework
SystemConfiguration.framework
CoreGraphics.framework
Foundation.framework
UIKit.framework
Security.framework
libz.dylib
```

(5) 设置 Search Paths 下的 Header Search Paths 和 Library Search Paths。若 SDK 文件夹(默认为 lib)与工程文件在同一级目录下，则都设置为“$(SRCROOT)/[文件夹名称]”，如图 9.35 所示。

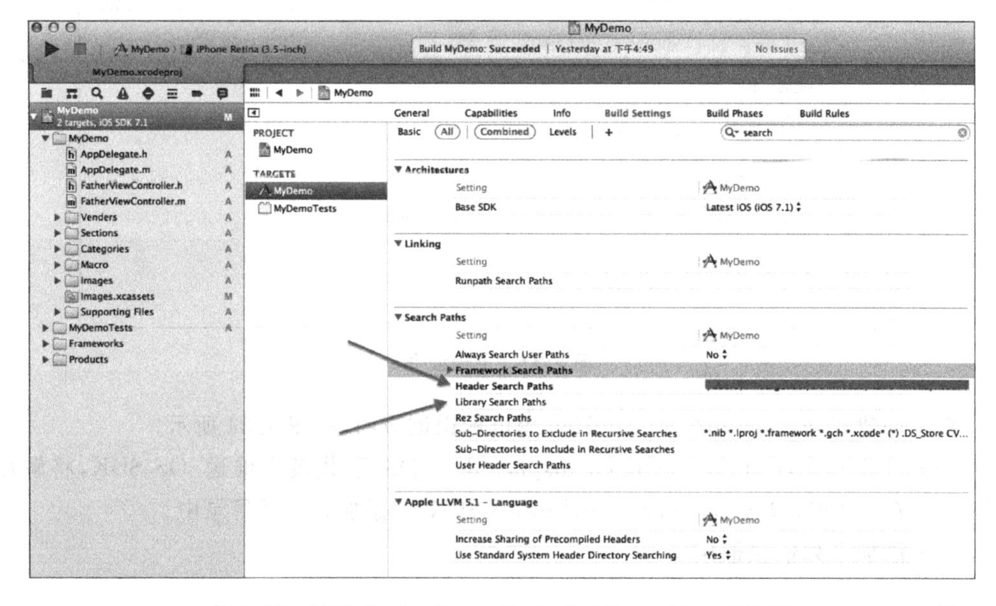

图 9.35　设置 Header Search Paths 和 Library Search Paths

(6) 在工程中创建一个新的 Property List 文件，并将其命名为 PushConfig. plist，如图 9.36 所示。

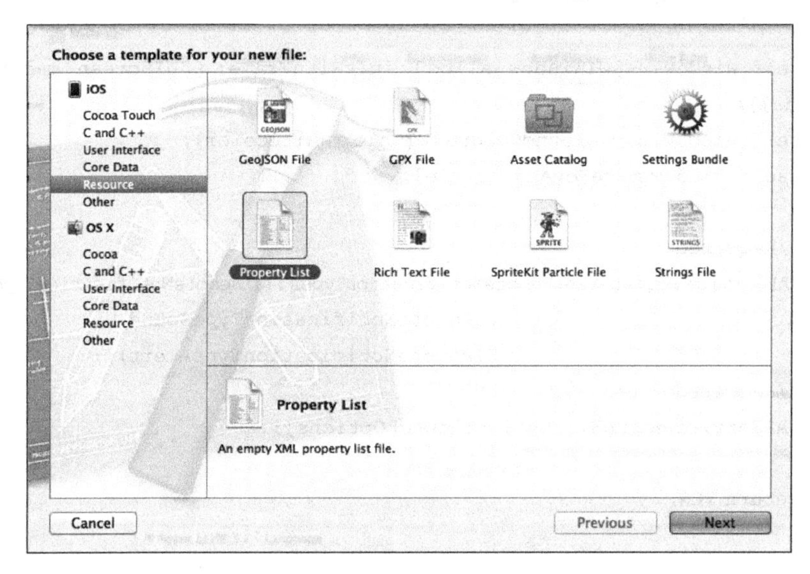

图 9.36 新建一个 Property List

(7) 单击 PushConfig. plist 文件，选择 Root 行的"+"号添加内容，如图 9.37 所示。

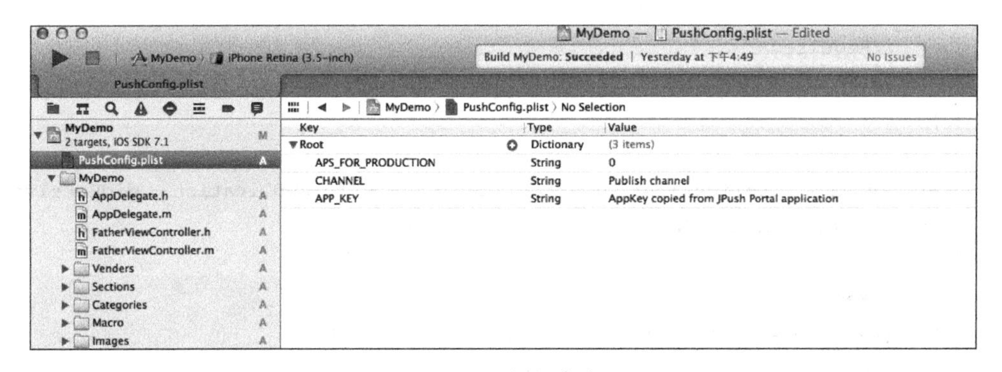

图 9.37 添加内容

说明如下：

CHANNEL——指明应用程序包的下载渠道，为方便分渠道统计，可根据需求自行定义。

App_KEY——在管理 Portal 上创建应用时自动生成(AppKey)，用以标识该应用。确保应用内配置的 AppKey 与在 Portal 上创建应用时生成的 AppKey 一致，AppKey 可以在应用详情中查询。

APS_FOR_PRODUCTION——应用是否采用生产证书发布(Ad_Hoc 或 App Store)，0(默认值)表示采用开发者证书，1 表示采用生产证书。

(8) 单击 AppDelegate. m，修改代码如下：

```
- (BOOL)application:(UIApplication *)application didFinishLaunchingWithOptions:
(NSDictionary *)launchOptions
{
    self.window=[[UIWindow alloc] initWithFrame:[[UIScreen mainScreen]
bounds]];
    self.window.backgroundColor=[UIColor whiteColor];
    [self.window makeKeyAndVisible];

    //Required
    [APService registerForRemoteNotificationTypes:(UIRemoteNotificationTypeBadge |
                            UIRemoteNotificationTypeSound |
                            UIRemoteNotificationTypeAlert)];
    //Required
    [APService setupWithOption:launchOptions];

    return YES;
}

- (void) application: (UIApplication *) application didRegisterForRemote
NotificationsWithDeviceToken:(NSData *)deviceToken {

    //Required
    [APService registerDeviceToken:deviceToken];
}

- (void) application: (UIApplication *) application didReceive
RemoteNotification:(NSDictionary *)userInfo {

    //Required
    [APService handleRemoteNotification:userInfo];
}
```

至此，极光推送就添加完成了。

9.6 基础知识与技能回顾

本章介绍了评价、分享和推送服务。现在网上出现了很多第三方框架供开发者选择，比如友盟的评价框架、百度的分享功能、极光的推送服务，这些第三方框架使用非常方便，大大减少了开发时间，应该多了解并运用这些第三方框架。

练 习

项目 1

功能描述：实现评分界面+功能。

项目 2

功能描述：上传 100 字以内的评价到服务端并显示。

项目 3

功能描述：实现分享商户信息到微信朋友圈。

第 10 章 chapter 10

添加商户信息

前面的章节中，我们介绍了登录、注册、定位、评论、分享等功能，但这些都是基于用户角度的功能。从商户的角度来看，最重要的功能应该是让商户自己添加商户信息到服务端，本章将介绍如何让商户上传信息到服务端并显示。

10.1 服务端接口的准备

本节将用到上传商户信息接口，下面给出详细信息。

接口地址：http://localhost:8080/meServer/shop.php?

调用方式：Post

返回数据格式：Json

上传商户信息接口的请求参数及说明如表 10.1 所示。

表 10.1 上传商户信息接口的请求参数

请求参数	必 选	类 型	说 明
act	Y	string	postShop
shopAddress	Y	string	商户地址
shopDesc	Y	string	商户描述
shopImage	Y	string	图片地址
shopLatitude	Y	string	商户纬度
shopLongitude	Y	string	商户经度
shopName	Y	string	商户名字
shopPhone	Y	string	商户电话
shopSpend	Y	string	商户人均花费
shopTime	Y	string	商户营业时间
shopType	Y	string	商户类型

上传商户信息接口的返回字段及说明如表 10.2 所示。

表 10.2 上传商户信息接口的返回字段

返回字段	字段类型	字段说明
flag	string	0：失败，1：成功
msg	string	信息说明

10.2 添加商户信息的实现

完成后的添加商户信息页面如图 10.1 所示。

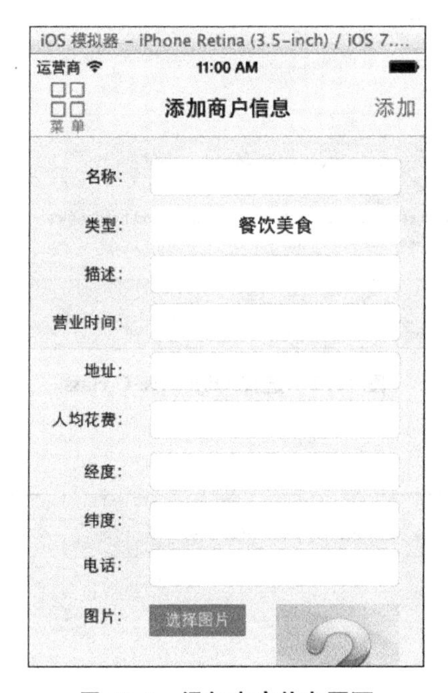

图 10.1 添加商户信息页面

10.2.1 客户端代码开发

1. 添加商户信息页面

(1) 单击 Sections 文件夹，按下 command+N，选择 Objective-C class，如图 10.2 所示。

(2) 在 Class 中填写 PostViewController，在 SubClass of 中填写 FatherViewController，不要选中 Also create XIB file，如图 10.3 所示。

(3) 单击 PostViewController.xib，先拖放一个 UIScrollView 控件到 xib 上，再拖放其他控件到 UIScrollView 上，如图 10.4 所示。

(4) 依次右击 xib 上的每个 UITextField 控件，设置 delegate 为 File's Owner，如图 10.5 所示。

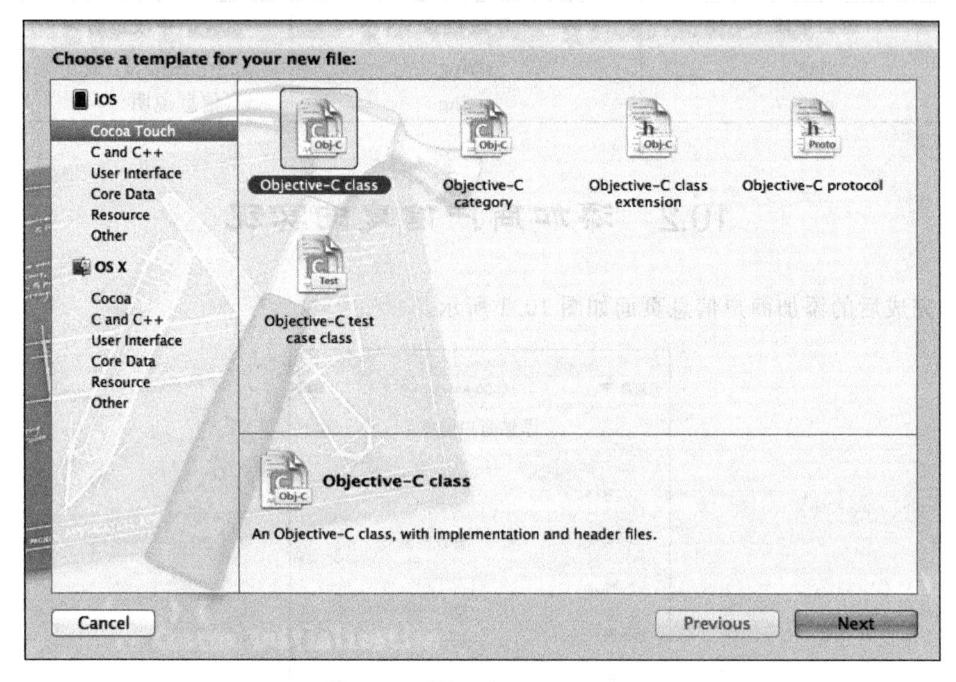

图 10.2　选择 Objective-C class

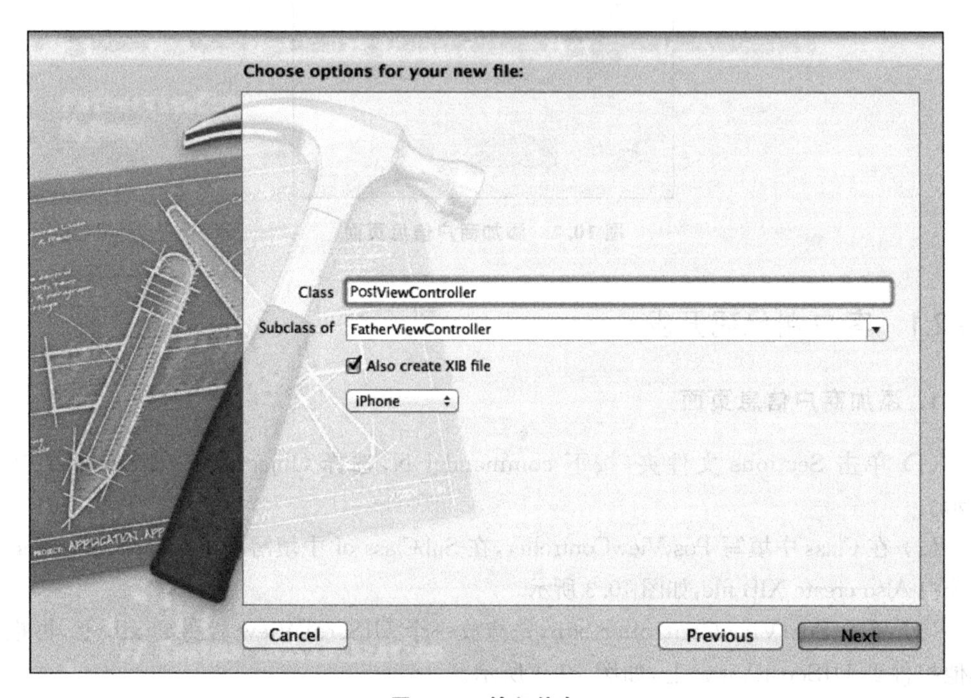

图 10.3　输入信息

图 10.4 拖放控件

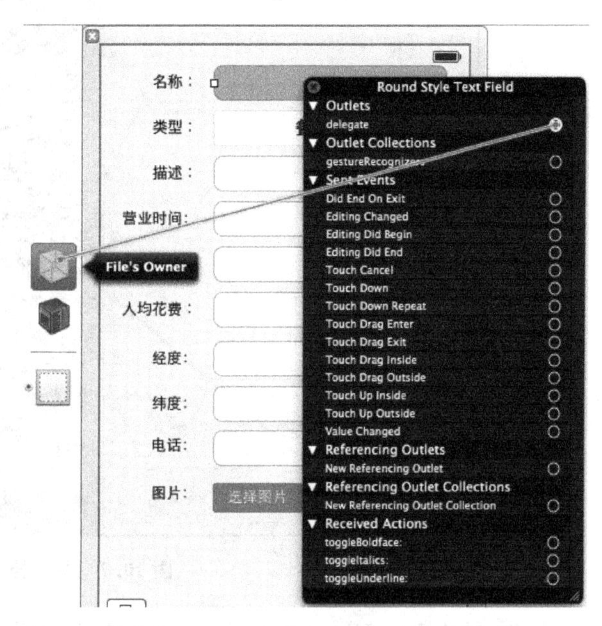

图 10.5 设置 UITextFie 的 delegate 属性为 File's Owner

(5) 依次设置 xib 上每个 UITextField 的 Tag 值，自上而下，第一个为 100，向下依次 +1，如图 10.6 所示。

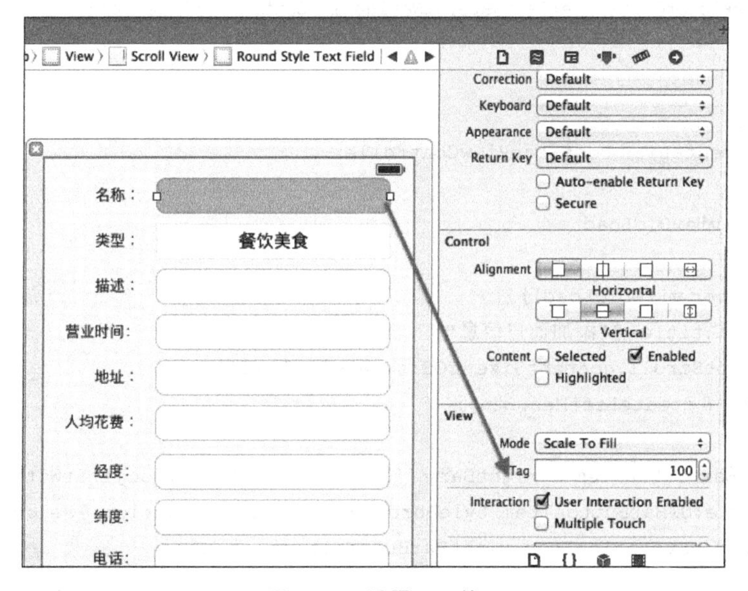

图 10.6 设置 Tag 值

(6) 关联“类型”按钮和“选择图片”按钮的 Touch Up InSide 事件，分别命名为 selectType 和 selectImage。关联 UIScrollView 和 UIImageView 控件的 New Referencing Outlet 属性，分别命名为 rootScroll 和 image，如图 10.7 所示。

图 10.7　关联控件

（7）单击 PostViewController.m 文件，添加代码如下：

```
#import "PostShopViewController.h"
@interface PostShopViewController ()<UIActionSheetDelegate,UITextFieldDelegate,
UIImagePickerControllerDelegate,UINavigationControllerDelegate>

@end

@implementation PostShopViewController

-(void)viewDidLoad
{
    [super viewDidLoad];
    self.title=@"添加商户信息";
    _rootScroll.contentSize=CGSizeMake(320, 480);
    [self createLeftMenuBtn];

    UIBarButtonItem * rightBar=[[UIBarButtonItem alloc]initWithTitle:@"添
加" style:UIBarButtonItemStyleBordered target:self action:@selector(add)];
    self.navigationItem.rightBarButtonItem=rightBar;
}
-(void)textFieldDidBeginEditing:(UITextField *)textField{
    CGRect frame=textField.frame;
    int offset=frame.origin.y+20-(self.view.frame.size.height-216.0);
                                                          //键盘高度 216
```

```
    NSTimeInterval animationDuration=0.1f;
    [UIView beginAnimations:@"ResizeForKeyboard" context:nil];
    [UIView setAnimationDuration:animationDuration];
    //将视图的Y坐标向上移动offset个单位,以使下面腾出地方用于软键盘的显示
    if(offset>0)
        self.view.frame = CGRectMake (0.0f, - offset, self.view.frame.size.
width, self.view.frame.size.height);
    [UIView commitAnimations];
}
- (void)textFieldDidEndEditing:(UITextField *)textField
{
    self.view.frame=CGRectMake(0, 64, self.view.frame.size.width, self.
view.frame.size.height);
}
- (BOOL)textFieldShouldReturn:(UITextField *)textField{
    [textField resignFirstResponder];
    return YES;
}
- (void)add{

}
- (void)didReceiveMemoryWarning
{
    [super didReceiveMemoryWarning];
    //Dispose of any resources that can be recreated.
}

- (IBAction)selectType:(id)sender {
    UIActionSheet * sheet=[[UIActionSheet alloc]initWithTitle:@"请选择商户
类型" delegate:self cancelButtonTitle:@"取消" destructiveButtonTitle:nil
otherButtonTitles:@"餐饮美食",@"休闲娱乐", nil];
    [sheet showInView:self.view];
}
- (IBAction)selectImage:(id)sender {
    UIImagePickerController * picker = [[UIImagePickerController alloc]
init];
    picker.delegate=self;
    picker.sourceType=UIImagePickerControllerSourceTypePhotoLibrary;
    picker.allowsEditing=YES;
    [self presentViewController:picker animated:YES completion:nil];


}
#pragma mark-UIImagePickerControllerDelegate
```

```
- ( void ) imagePickerController: ( UIImagePickerController * ) picker
didFinishPickingMediaWithInfo:(NSDictionary * )info
{
    UIImage * image= [info objectForKey:@"UIImagePickerControllerEditedImage"];
    _imageview.image= image;
    [picker dismissViewControllerAnimated:YES completion:^{}];
}

- (void) imagePickerControllerDIdCancel: (UIImagePickerController * )picker
{
    [picker dismissViewControllerAnimated:YES completion:nil];
}
@end
```

代码解析：

当单击 UITextField 完成输入后，弹出的键盘会遮挡界面，由于在介绍 xib 布局界面时，我们将所有 UITextField 的 delegate 属性设置为当前控制器，因此，只需实现 UITextFieldDelegate 协议的 textFieldShouldReturn：方法，并在其中添加[textField resignFirstResponder]；return YES；就可以了。按 return 键，键盘消失。

另外，在 textFieldDidBeginEditing：方法中，程序监听 UITextField 的编辑事件。当 UITextField 被编辑时，程序判断该 UITextField 是否被键盘遮挡。如被遮挡，则上移；如没被遮挡，则不做处理。

2. 为添加商户信息页面增加入口

(1) 单击 LeftViewController. xib，拖放一个 UIButton 到 xib，双击命名为“添加商户信息”，并关联 Touch Up Inside 事件，方法名为 goPost。拖放一个 UIimageView，Image 属性为 iconHomePageNormal. png，如图 10. 8 所示。

(2) 单击 LeftViewController. m，引入 PostViewController 头文件，代码如下：

```
#import "PostShopViewController.h"
```

重写 goPostVc：方法，代码如下：

```
- (IBAction)goPost:(id)sender {
    PostShopViewController * vc = [[ PostShopViewController alloc ]
initWithNibName:@"PostShopViewController" bundle:nil];
    UINavigationController * navigationController = self. menuContainer
ViewController.centerViewController;
    NSArray * controllers= [NSArray arrayWithObject:vc];
    navigationController.viewControllers=controllers;
    [self.menuContainerViewController setMenuState:MFSideMenuStateClosed];
}
```

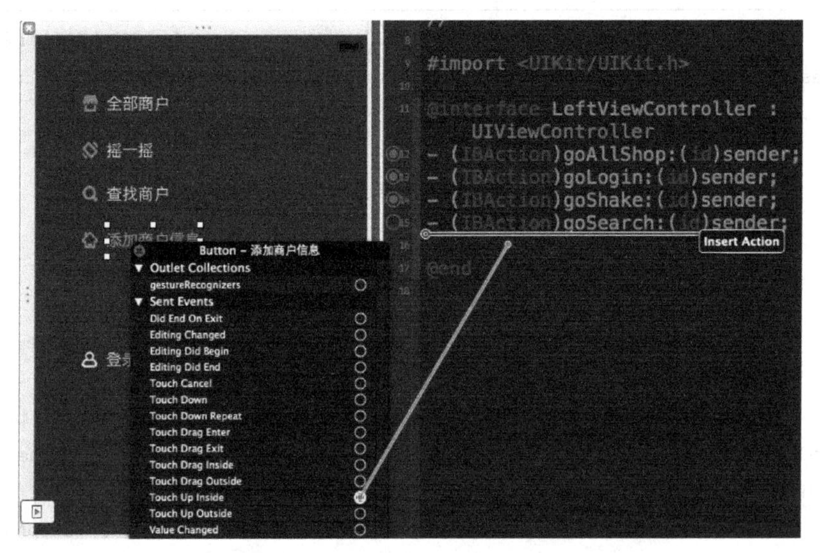

图 10.8 拖放控件并关联单击事件

10.2.2 客户端与服务端交互

(1) 单击 PostShopViewController.m，导入头文件，代码如下：

```
#import "AFHelper.h"
```

(2) 重写 done 方法，代码如下：

```
-(void)done{
    UITextField *shopNameTF=(UITextField *)[self.view viewWithTag:100];
    UITextField *shopDescTF=(UITextField *)[self.view viewWithTag:101];
    UITextField *shopTimeTF=(UITextField *)[self.view viewWithTag:102];
    UITextField *shopAddressTF=(UITextField *)[self.view viewWithTag:
103];
    UITextField *shopSpendTF=(UITextField *)[self.view viewWithTag:104];
    UITextField *shopLongitudeTF=(UITextField *)[self.view viewWithTag:
105];
    UITextField *shopLatitudeTF=(UITextField *)[self.view viewWithTag:
106];
    UITextField *shopPhoneTF=(UITextField *)[self.view viewWithTag:107];

    for(int i=0;i<8;i++){
        UITextField *textfield=(UITextField *)[self.view viewWithTag:100
+i];
        if(textfield.text.length<1){
```

```
            UIAlertView * alert= [[UIAlertView alloc] initWithTitle:@"提示" message:@"请完整输入" delegate:nil cancelButtonTitle:@"好的" otherButtonTitles:nil, nil];
            [alert show];
            return;
        }
    }
    if(_hasImage){
        NSData * imageData=UIImageJPEGRepresentation(_imageview.image, 0.5);
        NSDictionary * dic=@{@"act":@"postImage"};
        [AFHelper postImageWithDictionary: dic andImageData: imageData andImageName: @"2" andBaseURLStr: @"http://localhost: 8080/meServer/" andPostPath:@"image.php?" success:^(NSDictionary * dic){
            NSDictionary * dic1 = @ {@"act": @"postShop", @"shopAddress": shopAddressTF. text, @"shopDesc": shopDescTF. text, @"shopLatitude": shopLatitudeTF. text, @"shopLongitude": shopLongitudeTF. text, @"shopName": shopNameTF.text,@"shopPhone":shopPhoneTF.text,@"shopSpend":shopSpendTF.text, @"shopTime":shopTimeTF.text, @"shopType": [NSString stringWithFormat:@"%d", _shopType], @"shopImage": [NSString stringWithFormat: @"%@%@", @"http://localhost:8080/meServer",dic[@"image"]]};
            [AFHelper downDataWithDictionary: dic1 andBaseURLStr: @"http://localhost:8080/meServer/" andPostPath:@"shop.php?" success:^(NSDictionary * dic){
                UIAlertView *alert= [[UIAlertView alloc] initWithTitle:@"提示" message:dic[@"msg"] delegate:nil cancelButtonTitle:@"好的" otherButtonTitles:nil, nil];
                [alert show];
            }];
        }];
    }else{
        NSDictionary * dic = @ {@"act": @"postShop", @"shopAddress": shopAddressTF. text, @"shopDesc": shopDescTF. text, @"shopLatitude": shopLatitudeTF.text, @"shopLongitude": shopLongitudeTF. text, @"shopName": shopNameTF.text, @"shopPhone": shopPhoneTF.text, @"shopSpend": shopSpendTF. text,@"shopTime":shopTimeTF.text,@"shopType":[NSString stringWithFormat: @"%d",_shopType],@"shopImage":@"www.baidu.com"};
        [AFHelper downDataWithDictionary: dic andBaseURLStr: @"http://localhost:8080/meServer/" andPostPath:@"shop.php?" success:^(NSDictionary * dic){
```

```
            UIAlertView * alert = [[UIAlertView alloc] initWithTitle: @"提示"
message:dic[@"msg"] delegate:nil cancelButtonTitle:@"好的" otherButtonTitles:
nil, nil];
            [alert show];
        }];
    }

}
```

代码解析：

程序先判断用户是否完整输入。如没有完整输入，则提示用户完整输入。如已经完整输入，那么程序先判断是否有图片提交，有图片则先上传图片再提交商户信息，没有图片提交就直接上传商户信息。

10.3 基础知识与技能回顾

本章介绍了使用 App 上传商户信息到服务端，主要用 AFHelper 类实现对商户信息的上传。大家可以进行扩展，比如一次上传多张图片。

练 习

项目

功能描述：实现上传商户界面＋功能，并上传商户信息到服务端。

第11章 chapter 11

让用户的使用体验更佳

对 App 来说，用户体验如何将直接影响用户的参与度。如果 App 的用户体验非常糟糕，就不会通过苹果公司的审核。因此，在开发一款 App 时，良好的用户体验应该是重中之重。

11.1 用户网络环境

在 iOS 的开发中，检测用户网络环境非常重要。如果应用中不做网络环境变化的检测，很可能不会通过苹果公司的审核。苹果公司提供了一个类 Reachability 来检测网络环境的变化。

下载地址：

```
https://developer.apple.com/library/iOS/samplecode/Reachability/Reachability.zip
```

下载解压后将 Reachability.h 和 Reachability.m 添加到自己的项目中，并添加 SystemConfiguration.framework 框架，如图 11.1 所示。

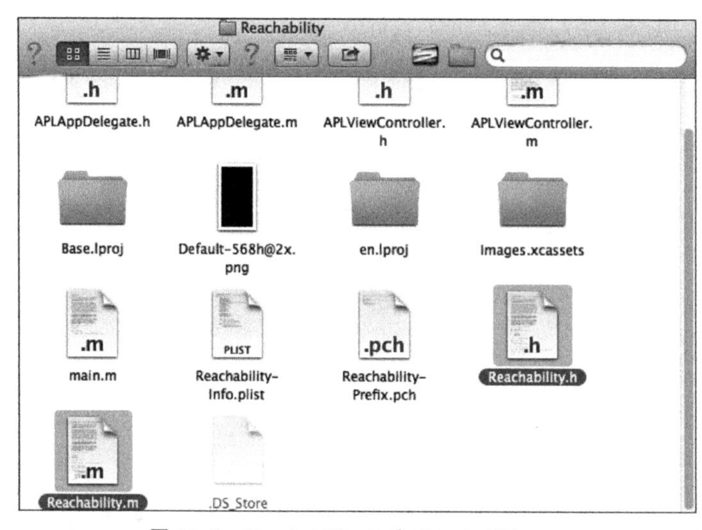

图 11.1　Reachability.h 和 Reachability.m

Reachability 中定义了 3 种网络状态：

```
//the network state of the device for Reachability 1.5. typedef enum {
    NotReachable=0,                      //无连接
    ReachableViaCarrierDataNetwork,      //使用 3G/GPRS 网络
    ReachableViaWiFiNetwork              //使用 WiFi 网络
} NetworkStatus;
  //the network state of the device for Reachability 2.0. typedef enum {
    NotReachable=0,                      //无连接
    ReachableViaWiFi,                    //使用 3G/GPRS 网络
    ReachableViaWWAN                     //使用 WiFi 网络
} NetworkStatus;
```

比如检测某一特定站点的连接状况，可以使用下面的代码：

```
Reachability * r=[Reachability reachabilityWithHostName:@“www.apple.com”];
  switch ([r currentReachabilityStatus]) {
    case NotReachable:                   //没有网络连接
        break;
    case ReachableViaWWAN:
        //使用 3G 网络
        break;
    case ReachableViaWiFi:
        //使用 WiFi 网络
        break;
}
```

检测当前网络环境，代码如下：

```
//是否 WiFi+(BOOL) IsEnableWIFI {
    return ([[Reachability reachabilityForLocalWiFi] currentReachabilityStatus]
!=NotReachable);
}
//是否 3G
+(BOOL) IsEnable3G {
    return  ([[ Reachability  reachabilityForInternetConnection ]  current
ReachabilityStatus] !=NotReachable);
}
```

网络连接状态的实时检查、通知在网络应用中也是十分必要的。连接状态发生变化时，需要及时通知用户。这里以 Reachability 2.0 版为例来说明使用方法，代码如下：

```
//AppDelegate.h
@class Reachability;
@interface AppDelegate: NSObject<UIApplicationDelegate>{
    Reachability  *hostReach;
}
@end
//AppDelegate.m
-(void)reachabilityChanged:(NSNotification *)note {
    Reachability* curReach=[note object];
    NSParameterAssert([curReach isKindOfClass: [Reachability class]]);
    NetworkStatus status=[curReach currentReachabilityStatus];
    if (status==NotReachable) {
        UIAlertView *alert=[[UIAlertView alloc] initWithTitle:@"AppName""
                            message:@"NotReachable"delegate:nil
                            cancelButton Title:"YES" otherButton Titles:nil];
        [alert show];
    }
  }
-(void)applicationDidFinishLaunching:(UIApplication *)application {
    //...
      //监测网络情况
     [[NSNotificationCenter defaultCenter] addObserver: self selector: @
selector (reachabilityChanged:) name: kReachabilityChangedNotification
object: nil];
    hostReach=[Reachability reachabilityWithHostName:@"www.google.com];
    [hostReach startNotifer];
    //...
}
```

11.2 用户手机环境

从第一代 iPhone 到如今的 iPhone5S/iPhone5C，苹果公司已经推出了 7 代 iPhone。iPhone 3G 到 iPhone 5S/iPhone 5C 的屏幕信息如表 11.1 所示。

表 11.1 iPhone3G 到 iPhone5S/iPhone5C 的屏幕信息

型　号	屏幕尺寸	分 辨 率
iPhone 3G	3.5 英寸	480×320 像素
iPhone 3GS	3.5 英寸	480×320 像素
iPhone 4	3.5 英寸	960×640 像素
iPhone 4S	3.5 英寸	960×640 像素
iPhone 5	4 英寸	1136×640 像素
iPhone 5S/iPhone 5C	4 英寸	1136×640 像素

可以看出，iPhone 的屏幕大小并不统一，因此需要对不同 iPhone 的设备进行适配。

1. 动态获取设备属性

在前面章节中介绍了宏定义，比如：

```
#define App_SCREEN_WIDTH  [UIScreen mainScreen].bounds.size.width
#define App_SCREEN_HEIGHT  [UIScreen mainScreen].bounds.size.height
#define App_SCREEN_CONTENT_HEIGHT  ([UIScreen mainScreen].bounds.size.
height-STATUEBAR_HEIGHT)
#define IOS_7  ([[[UIDevice currentDevice] systemVersion]floatValue]>=7.0?
YES:NO)
#define IS_4_INCH  (App_SCREEN_HEIGHT>480.0)
```

在 iPhone 5 以上的设备中，App_SCREEN_HEIGHT 的值为 568，在 iPhone 5 以下设备中，App_SCREEN_HEIGHT 的值为 480. 因此在设置 Frame 属性时，要结合 App_SCREEN_HEIGHT 等宏定义去动态获取设备的属性。

2. autoresizingMask 属性

UIView 有一个非常重要的属性：autoresizingMask。它表示父视图的 frame 变化后，它在父视图中的位置或大小如何变化。

属性说明如下：

```
enum {
UIViewAutoresizingNone                 =0,
UIViewAutoresizingFlexibleLeftMargin   =1<<0,
UIViewAutoresizingFlexibleWidth        =1<<1,
UIViewAutoresizingFlexibleRightMargin  =1<<2,
UIViewAutoresizingFlexibleTopMargin    =1<<3,
UIViewAutoresizingFlexibleHeight       =1<<4,
UIViewAutoresizingFlexibleBottomMargin =1<<5
};
typedef NSUInteger UIViewAutoresizing;
```

假如背景中有一个 UIButton，我们希望这个 UIButton 无论什么时候距离其背景底部的高度都是固定的，那么只需要设置 UIButton 的 autoresizingMask 为 UIView Autoresi-zingFlexibleTopMargin，就能达到这个目的。

3. iOS7 导航栏

在 iOS7 中，如果希望在导航栏中使用一个图片作为背景，那么需要提供两种高度的图片(这样可以延伸到导航栏背后)。导航栏的高度从 44 points(88 pixels)变成了 64 points(128 pixels)。

11.3 基础知识与技能回顾

本章介绍了如何检测 iOS 设备的网络环境和适配 iOS 设备。

当 iOS 设备的网络环境发生改变时，程序需要友好地提示用户，并引导用户完成相应操作，使程序具有良好的交互性。

与 Android 开发者相比，iOS 开发者是幸福的，因为 iOS 开发者不用面对尺寸纷乱复杂的 Android 设备。在程序开发完成后，最好能在不同设备上测试，完善适配。

练　　习

项目

功能描述：用 Reachability 检测设备当前的网络环境。

第 12 章 chapter 12

发布和管理 iOS 应用

App 开发完成后，需要提交到 App Store 审核应用。审核通过后，App 就会出现在 App Store 中，用户可以正常下载使用。另外，我们也要管理好 App 的源代码，做好项目版本控制，方便新版本的发布和代码的回滚。

12.1 发布 iOS 应用

要发布 iOS 应用程序，目前有三种方式：

(1) 发布到 App Store。需要注册开发者账号，账号是收费的，99 $ /年。

(2) 发布到企业内部用户。需要注册企业账号，299 $ /年。企业账号通过 iTunes 同步、iPhone 配置工具及建立私有的应用安装网站等方式进行发布，不需要经过 App Store 审核，当然也不能提交到 App Store 中。

(3) Ad Hoc 发布。针对测试设备，每个应用安装不能超过 100 个设备，发布前需要将每个设备的唯一编码打包进去。

下面介绍最主流也是最重要的方式，如何将应用发布到 App Store。

12.1.1 申请发布证书

(1) 登录 https://developer.apple.com，选择 iOS Developer Program 下的 Certificates，Identifiers & Profiles，如图 12.1 所示。

(2) 单击 Certificates，如图 12.2 所示。

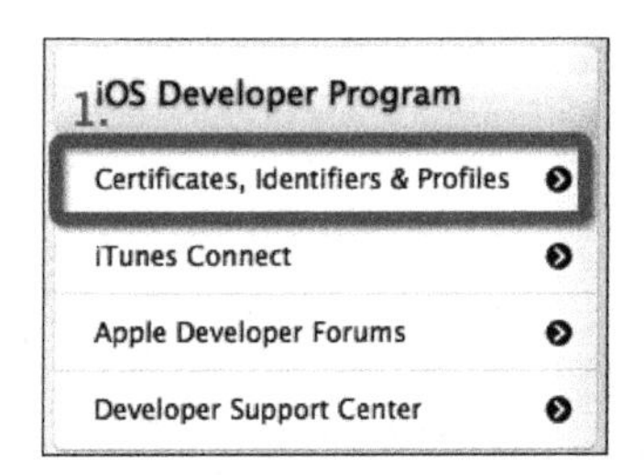

图 12.1 选择 Certificates，Identifiers & Profiles

图 12.2 单击 Certificates

(3) 选择证书类型 Distribution,选择添加,如图 12.3 所示。

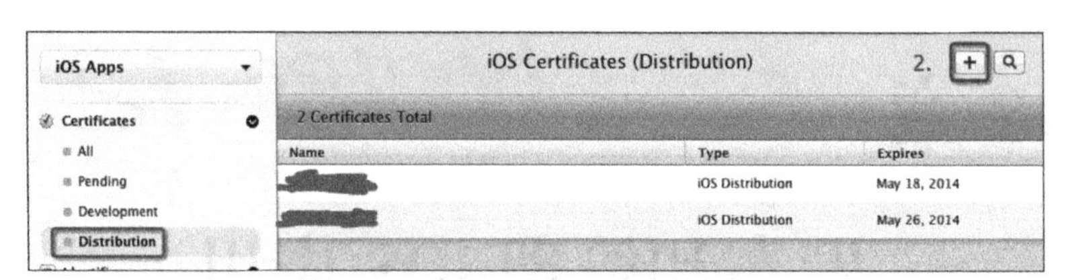

图 12.3 选择 Distribution

(4) 单击"+",选择 Distribution 下的 App Store and Ad Hoc,单击 Continue。

现在的 iTunes 已经将这两项分开,要上传至 App Store 请选择 App Store。如果要安装到一台或多台设备上,请选择 Ad Hoc,如图 12.4 所示。

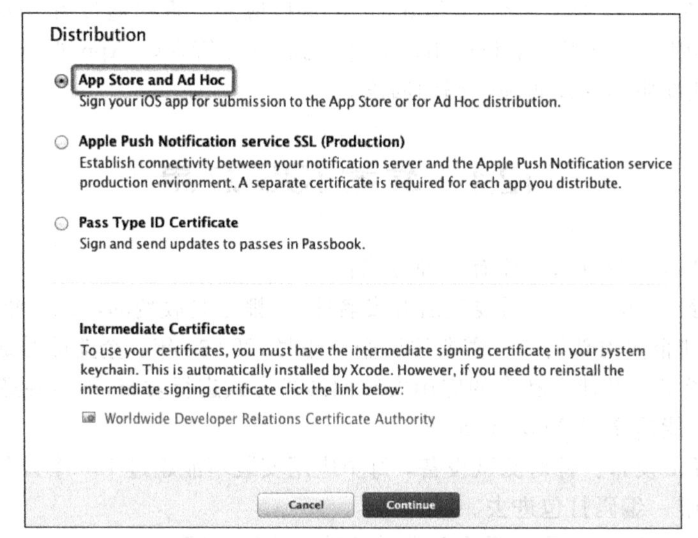

图 12.4 选择 App Store and Ad Hoc

(5) 单击 Continue,会看到需要上传证书界面,如图 12.5 所示。

图 12.5 上传证书界面

(6) 打开 Mac 的钥匙串访问,如图 12.6 所示。

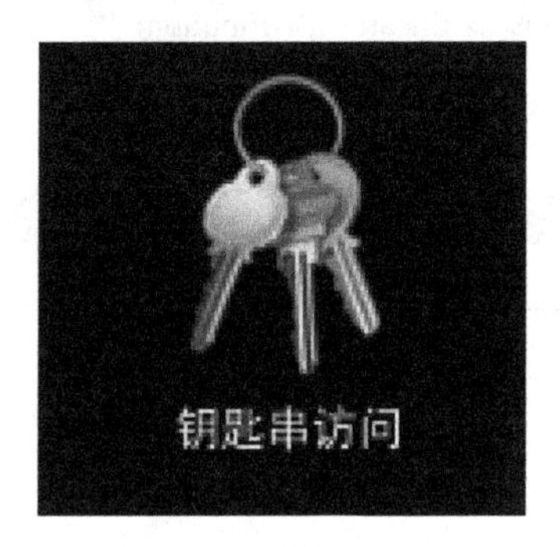

图 12.6　钥匙串访问

(7) 选择钥匙串的证书助理,如图 12.7 所示。

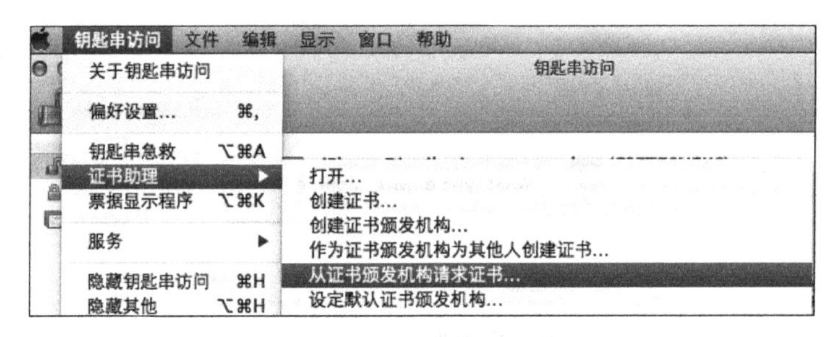

图 12.7　选择证书助理

(8) 单击“继续”,存储证书,如图 12.8 所示。

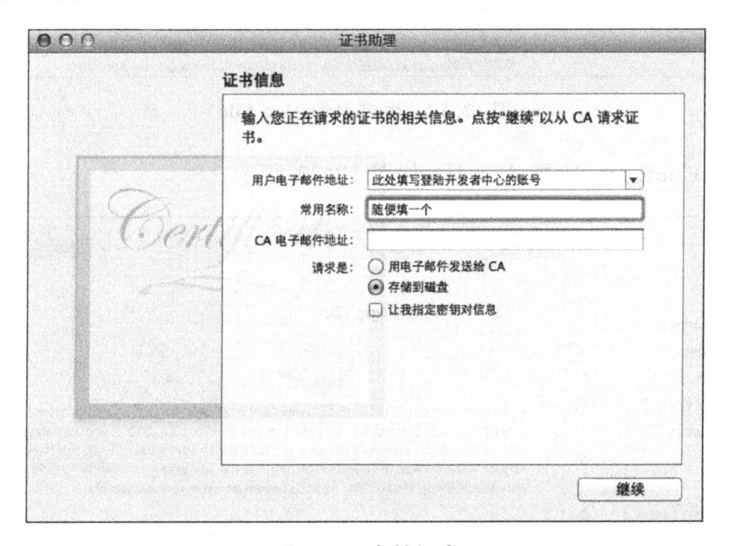

图 12.8　存储证书

(9) 在步骤(5)的界面选择 Choose File,将刚刚存储的证书上传。

(10) 下载证书,双击安装,如图 12.9 所示。

(11) 生成证书对应的 Provision File,如图 12.10 所示。

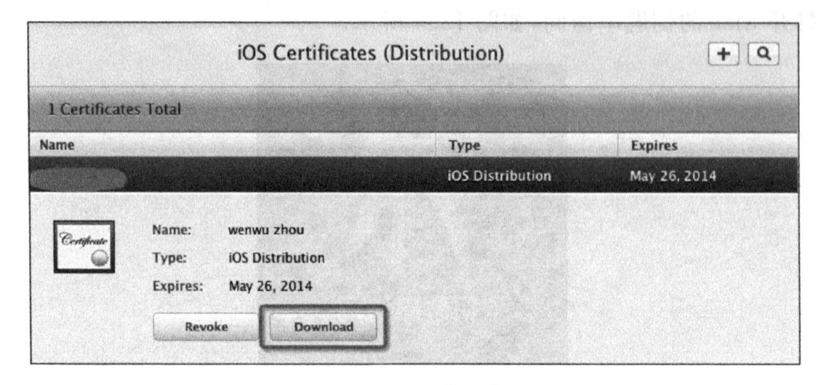

图 12.9 下载证书

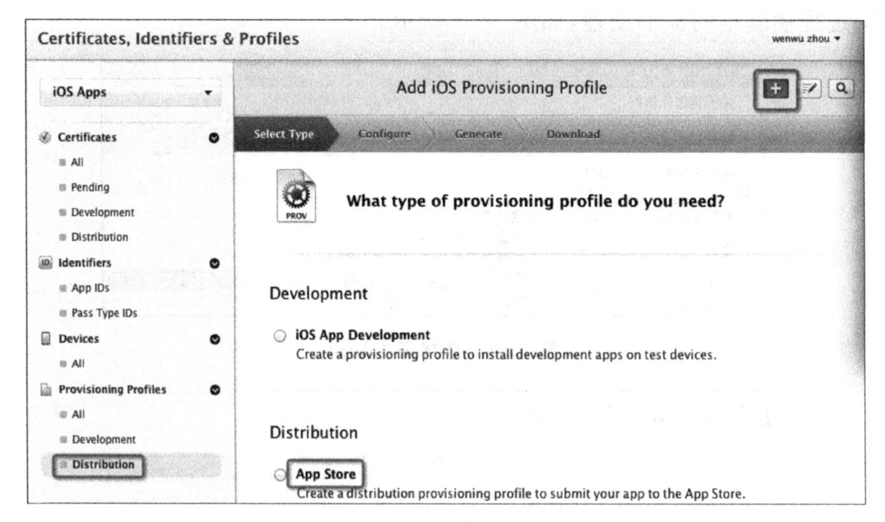

图 12.10 生成 Provision File

(12) 单击 Continue,选择 App ID,如图 12.11 所示。

图 12.11 选择 App ID

此处的 App ID 应选择自己原来建的,应该在真机调试时就生成过。如果没有,则到 Identifiers 下的 App IDs 中新建一个。

(13) 单击 Continue，选择刚刚新建的发布证书，如图 12.12 所示。

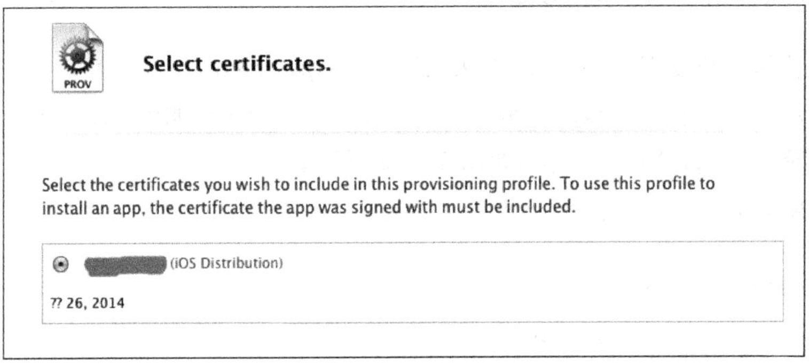

图 12.12 选择证书

(14) 选择 Continue，输入 Profile Name，如图 12.13 所示。

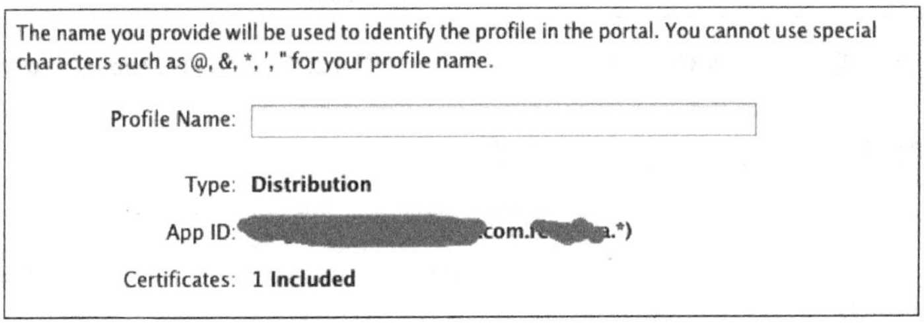

图 12.13 输入 Profile Name

此处填写的 Profile Name 应与 App ID 的后缀一致。如果为 *，则自定义一个。

(15) 单击生成后会在 Provision Profiles 里看到生成的 Profiles。选择 Type 为 Distribution 的项并下载，下载完成后双击安装，发布证书就申请并安装完成了，如图 12.14 所示。

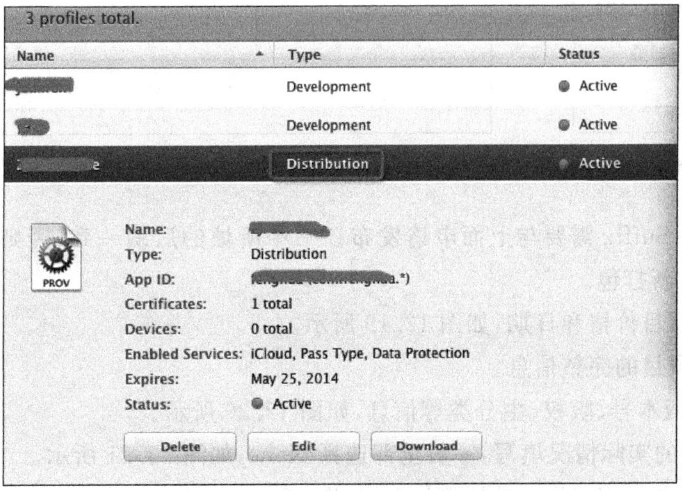

图 12.14 选择证书下载并安装

12.1.2 发布应用到 App Store

(1) 登录苹果开发者中心 http://developer.apple.com。

选择 iOS Developer Program 板块下的 iTunes Connect，如图 12.15 所示。

(2) 选择 Manage Your Apps，如图 12.16 所示。

(3) 选择 Add New App，如图 12.17 所示。

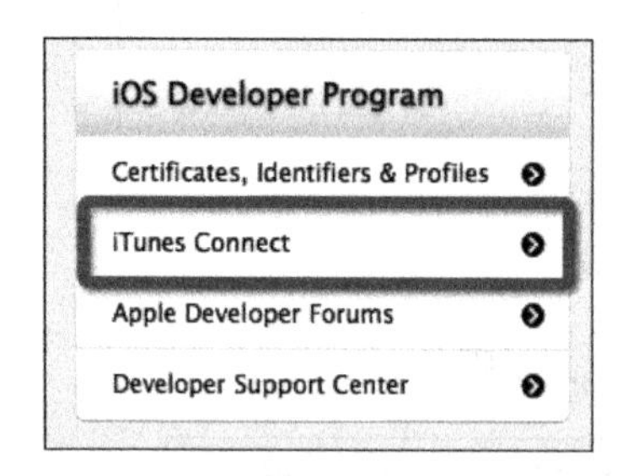

图 12.15 选择 iTunes Connect

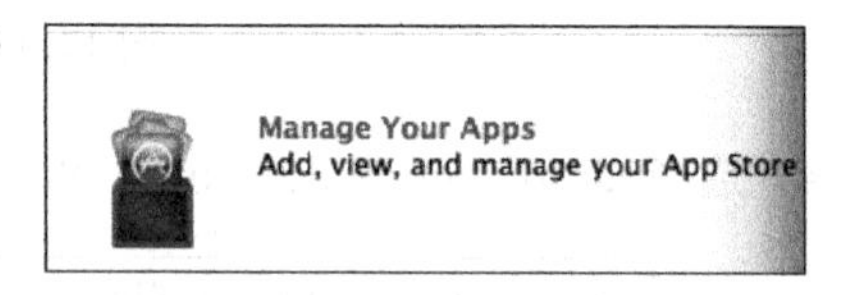

图 12.16 选择 Manage Your Apps

(4) 填写项目相关信息，如图 12.18 所示。

图 12.17 选择 Add New App

图 12.18 填写信息

Bundle ID Suffix 需要与上面申请发布证书中所填的后缀一致，否则在上传项目时会出错，需要重新打包。

(5) 选择项目价格和日期，如图 12.19 所示。

(6) 填写项目的完整信息。

依次填写版本号、版权、主分类等信息，如图 12.20 所示。

根据 App 的实际情况填写，一般全部选择 None，如图 12.21 所示。

填写描述、关键字、URL 地址等信息，如图 12.22 所示。

Select the availability date and price tier for your app.

Availability Date 05/May 26 2013

Price Tier Free

View Pricing Matrix ▶

Discount for Educational Institutions ☑

Unless you select specific stores, your app will be for sale in all App Stores worldwide.

图 12.19 选择项目价格和日期

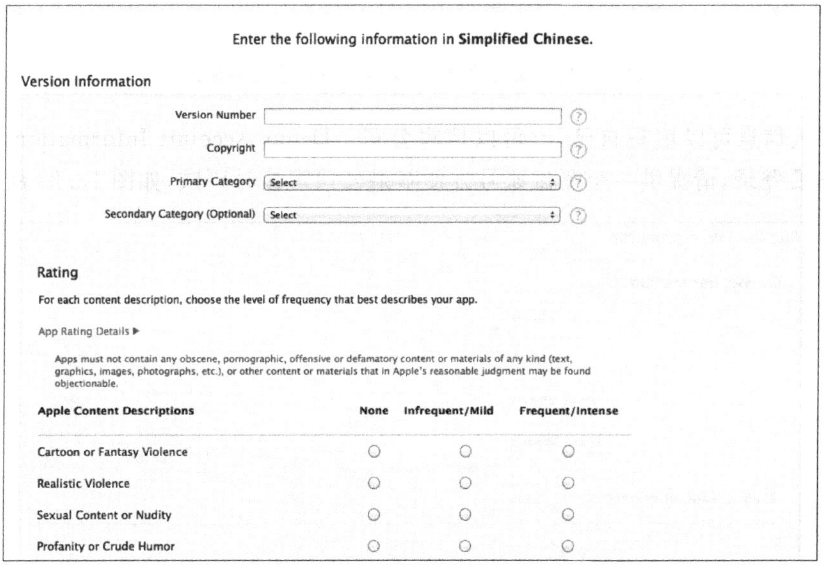

Enter the following information in **Simplified Chinese.**

Version Information

Version Number

Copyright

Primary Category Select

Secondary Category (Optional) Select

Rating

For each content description, choose the level of frequency that best describes your app.

App Rating Details ▶

Apps must not contain any obscene, pornographic, offensive or defamatory content or materials of any kind (text, graphics, images, photographs, etc.), or other content or materials that in Apple's reasonable judgment may be found objectionable.

Apple Content Descriptions	None	Infrequent/Mild	Frequent/Intense
Cartoon or Fantasy Violence	○	○	○
Realistic Violence	○	○	○
Sexual Content or Nudity	○	○	○
Profanity or Crude Humor	○	○	○

图 12.20 填写版本号、版权、主分类等信息

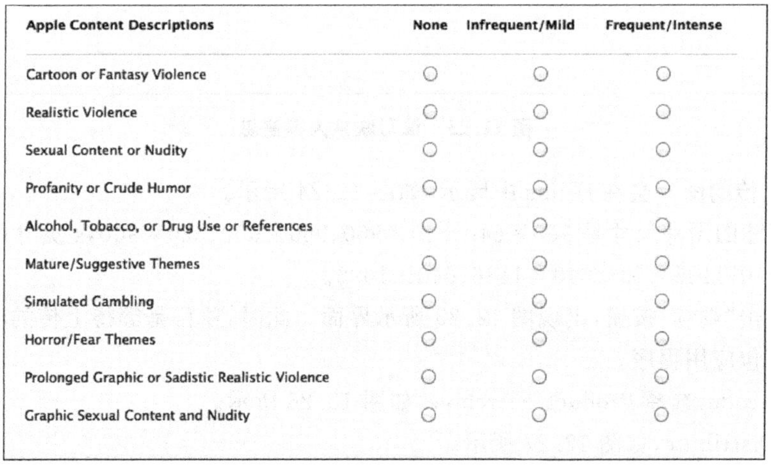

Apple Content Descriptions	None	Infrequent/Mild	Frequent/Intense
Cartoon or Fantasy Violence	○	○	○
Realistic Violence	○	○	○
Sexual Content or Nudity	○	○	○
Profanity or Crude Humor	○	○	○
Alcohol, Tobacco, or Drug Use or References	○	○	○
Mature/Suggestive Themes	○	○	○
Simulated Gambling	○	○	○
Horror/Fear Themes	○	○	○
Prolonged Graphic or Sadistic Realistic Violence	○	○	○
Graphic Sexual Content and Nudity	○	○	○

图 12.21 根据 App 的实际情况填写

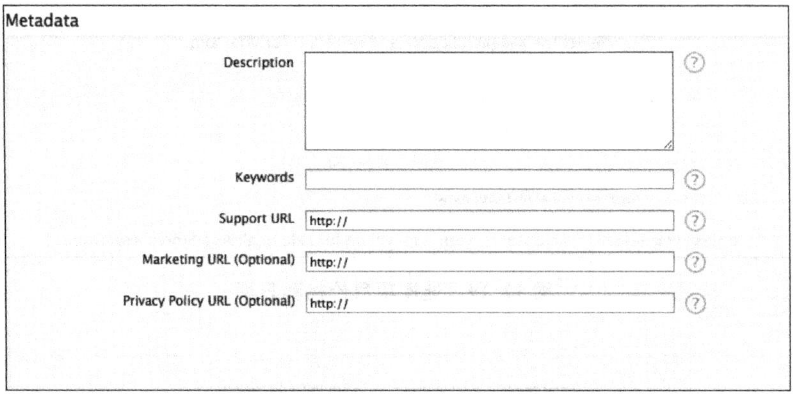

Metadata

Description

Keywords

Support URL http://

Marketing URL (Optional) http://

Privacy Policy URL (Optional) http://

图 12.22 填写描述、关键字、URL 地址等信息

联系人信息可以填写自己,也可以填写公司。Demo Account Information 指:如果 App 中需要登录,请提供一个测试账号方便苹果公司审核时调试,如图 12.23 所示。

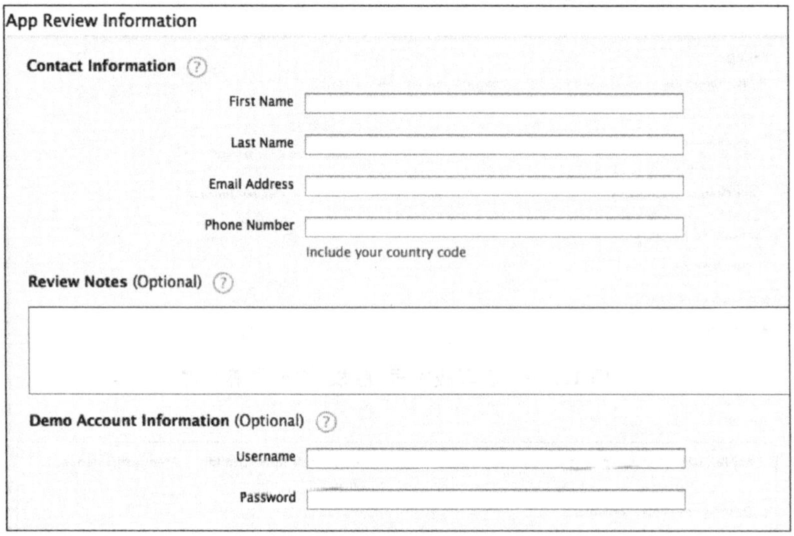

App Review Information

Contact Information

First Name

Last Name

Email Address

Phone Number

Include your country code

Review Notes (Optional)

Demo Account Information (Optional)

Username

Password

图 12.23 填写联系人等信息

此处上传的图片会在 iTuns 中展示,如图 12.24 所示。

3.5 英寸的图片尺寸是 960×640、640×960、960×600、620×900,4 英寸的图片尺寸是 1136×640、1136×600、640×1136、640×1096。

(7) 单击“继续”按钮,出现图 12.25 所示界面。此时,项目是等待上传的状态。

(8) 打包应用程序。

打开 Xcode,选择 Product→Archive,如图 12.26 所示。

选择 Distribute,如图 12.27 所示。

选择第二个选项,如图 12.28 所示。

选择发布证书,单击 Next 之后便会生成.ipa 文件,如图 12.29 所示。

图 12.24 上传图片

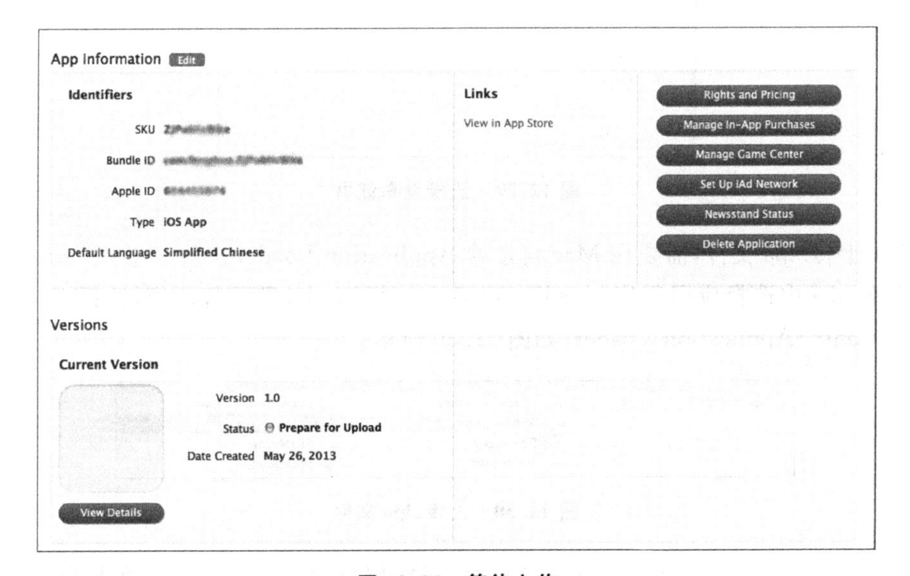

图 12.25 等待上传

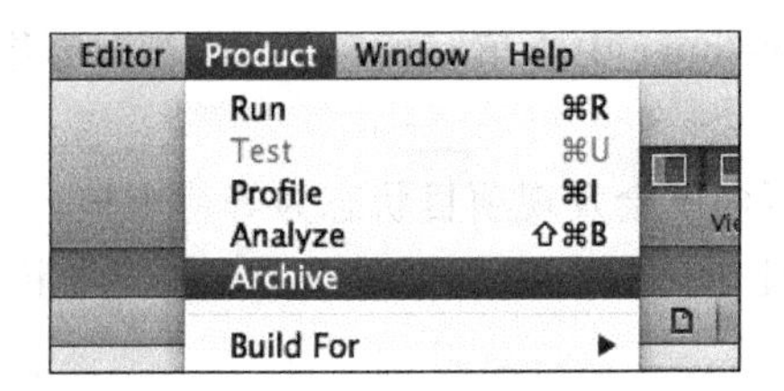

图 12.26 打包应用程序

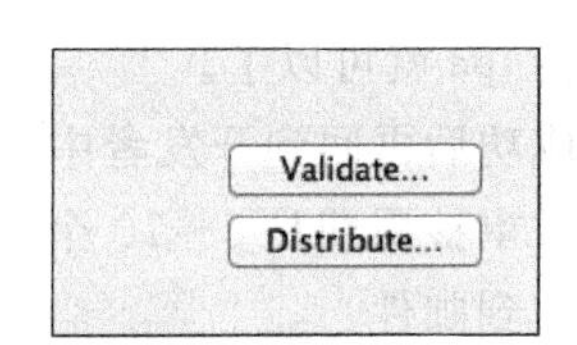

图 12.27 选择 Distribute

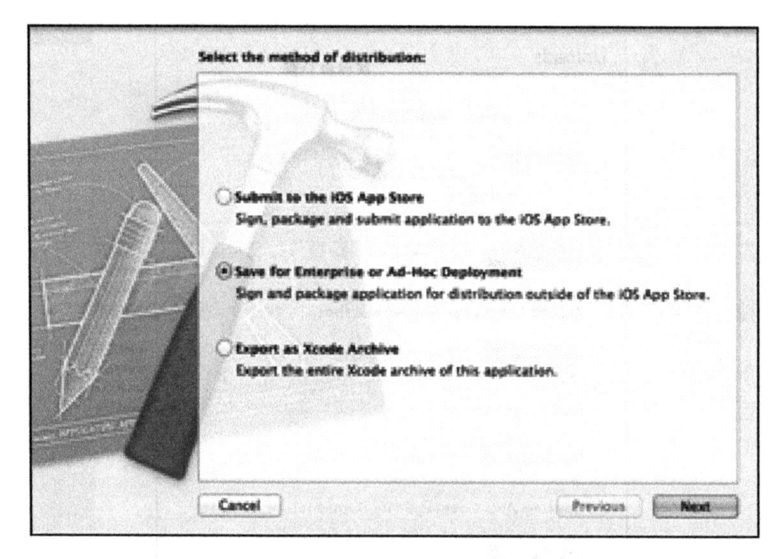

图 12.28 选择第二个选项

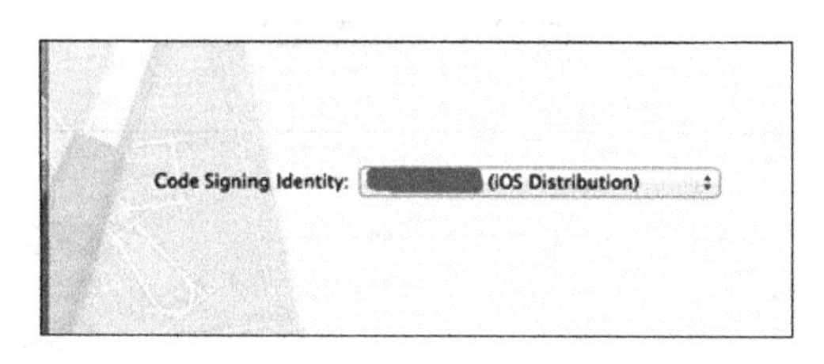

图 12.29 选择发布证书

(9) 上传.ipa 文件,需要用 Mac 自带的 Application Loader。

在 Finder 中选择应用程序(Applications)。右击 Xcode,显示包内容,选择 Contents→Applications→Application Loader,如图 12.30 所示。

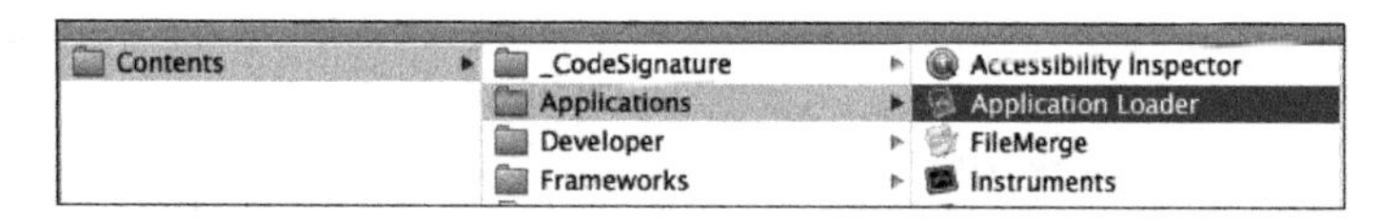

图 12.30 上传、ipa 文件

打开 Application Loader 出现登录界面,同样用登录开发者中心的账号登录。第一次登录会出现一个接受协议页面,选择接受后单击继续。

选择 Deliver Your App,会出现在 Developer 中心填写的项目,状态为等待上传,这里选择上传 ipa 就可以了。

上传成功后再回到开发者中心,刷新页面就会发现项目状态为等待审核,如图 12.31 所示。正常审核周期是 5~10 个工作日,此时邮箱会收到苹果公司的邮件,审核结果出来后也会收到邮件。

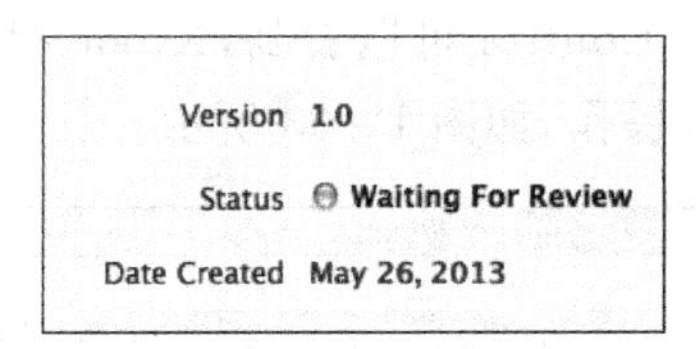

图 12.31 等待审核

12.2 版本管理

在开发过程中经常会遇到这样的问题，即本来编译、运行都没有问题的程序，不知修改了哪个配置，突然变得无法编译，或者运行时出现了各种各样的错误，此时只能是硬着头皮去一个个地修改，或者干脆全部推倒重来。这种方式面临很严重的问题，即使项目很小，也会耗费开发者的很多精力。如果项目规模庞大，那就彻底不知所措了。因此，在代码中实行版本控制是非常重要的。常用的版本控制工具有 SVN 和 GIT 客户端等，它在开发 iOS 应用时可以发挥强大的作用。如果只是一个自己开发的工程，可不必这么复杂。iOS 的 IDE 已经集成了一个简易的 Git，可以很方便地实现本地代码版本控制。

1. 为代码创建代码版本控制

(1) 为工程创建版本控制非常简单。只需在创建工程时在图示的位置勾选就可以了，如图 12.32 所示。

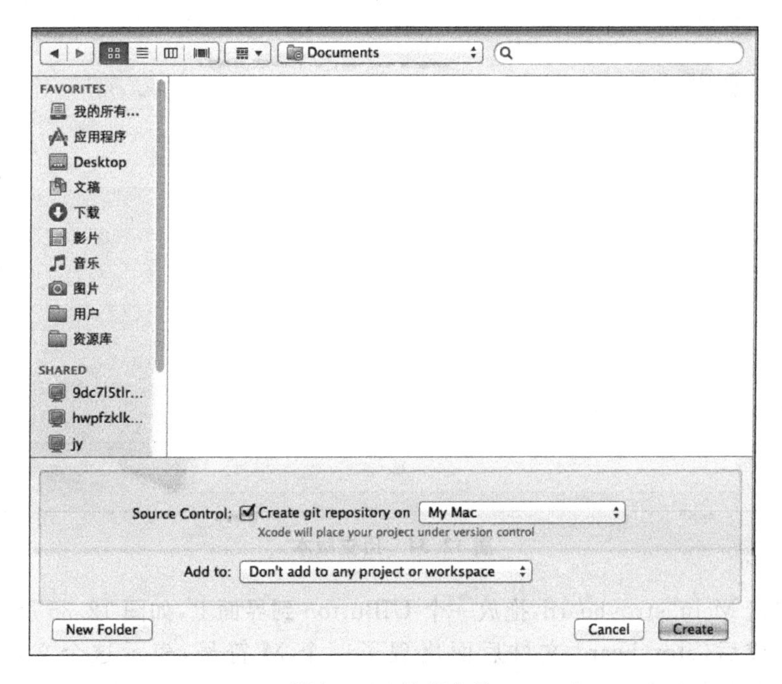

图 12.32 注意勾选

(2) 在菜单中选择 Source Control 可以看到，Xcode 提供的版本控制工具已经比较强大，支持分支的建立、合并等操作，如图 12.33 所示。

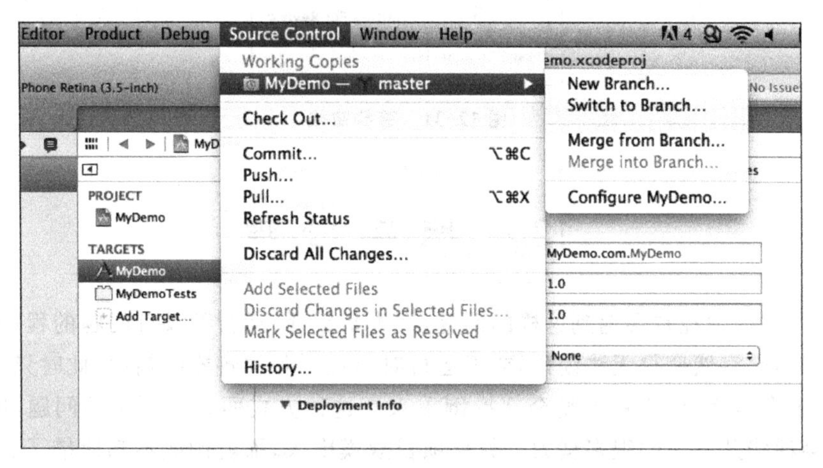

图 12.33　选择 Source Control

2. Xcode 中向版本控制工具提交修改

(1) 工程建立完成后，选择 Source Control→History，可以查看版本修订的历史。由于现在刚建好工程，Project history 窗口中只有一个 Initial Commit，表示第一次向服务端(其实就是本机)提交代码，如图 12.34 所示。

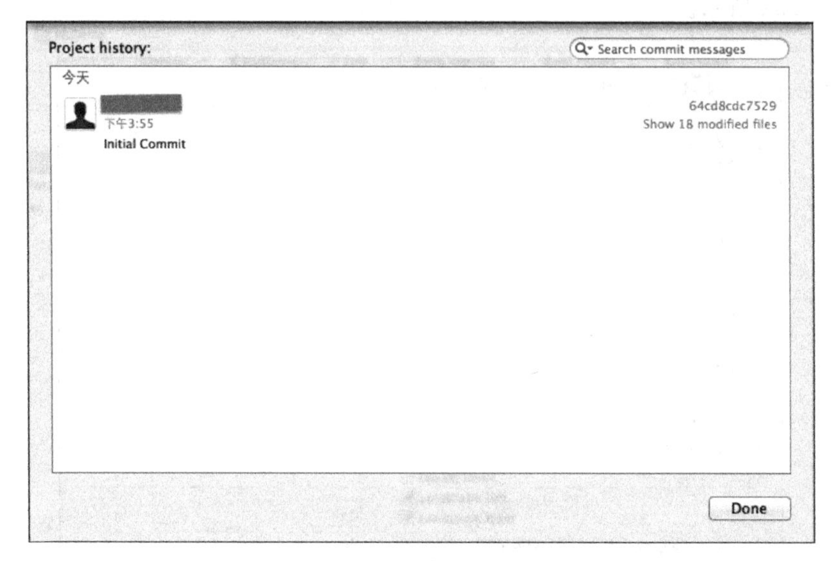

图 12.34　历史版本

(2) 单击 Main. stroyboard，拖放一个 UIButton 到界面上，如图 12.35 所示。

(3) 完成后，storyboard 文件后面出现了一个 M 符号，表示这个文件被修改过(Modified)但尚未提交。此时选择 Source Control→Commit，出现提交窗口。在这个窗

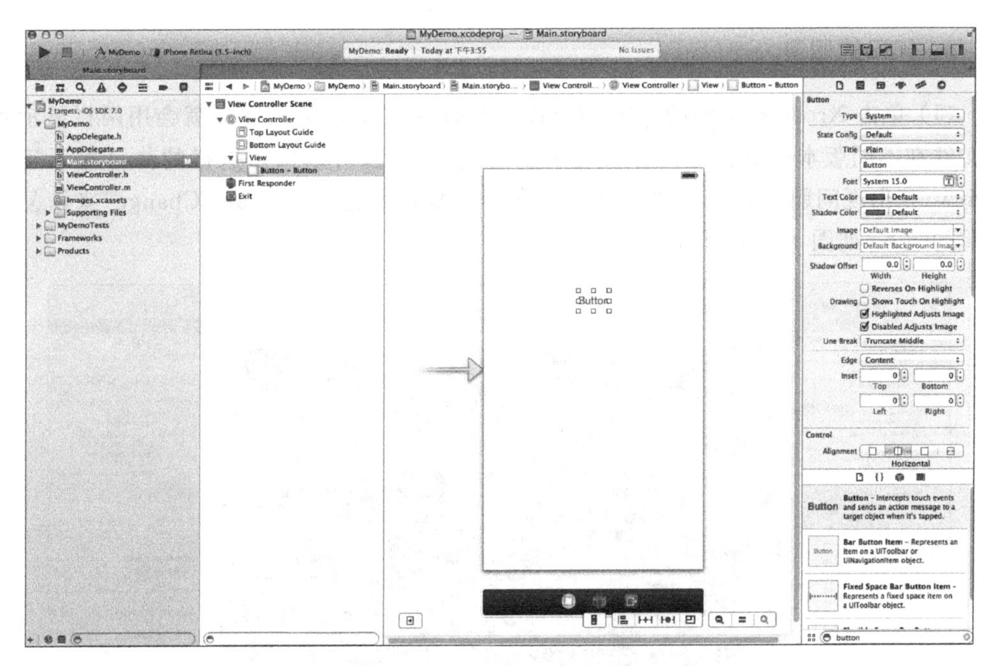

图 12.35　拖放 UIButton 到界面上

口里可以检查做了哪些修改，可选择性地提交修改，如图 12.36 所示。

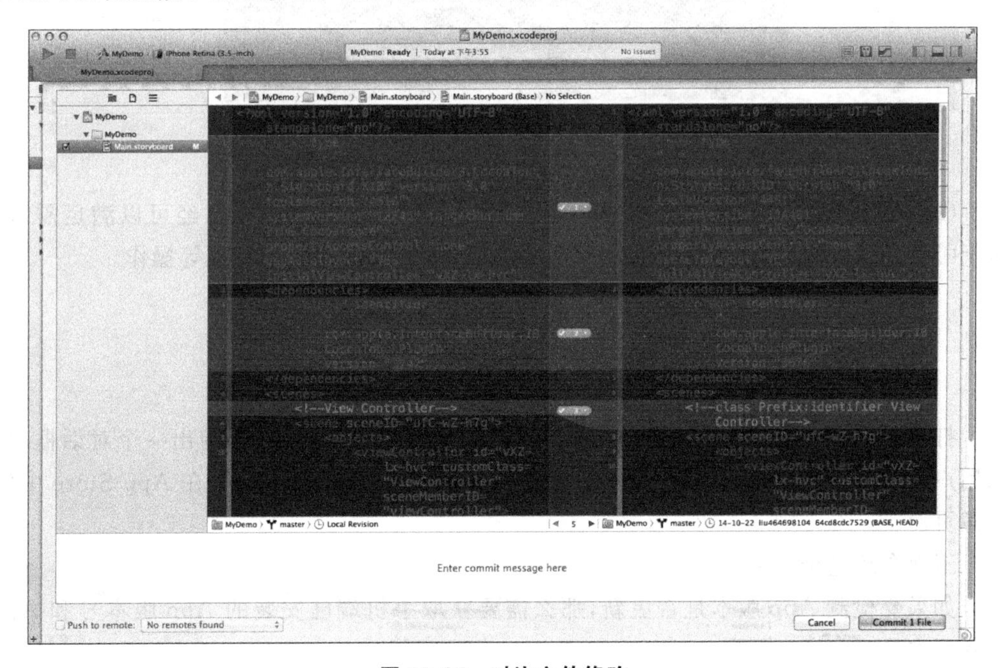

图 12.36　对比文件修改

(4) 单击 Commit 1 File 提交修改，在下面的 Enter commit message here 中写上对本次修改的说明。完成后再查看 History，就会看到提交记录了。

3. 代码的回退

(1) 单击 Xcode 右上角的按钮可以打开版本编辑器，之后界面就会用两栏显示各个文件在不同版本中的修订。在代码出现问题的时候，可以按照这样的修订提示逐条review，在出现问题需要回退的地方可以直接在修订处选择“Discard Change”来恢复到以前的版本，如图 12.37 所示。

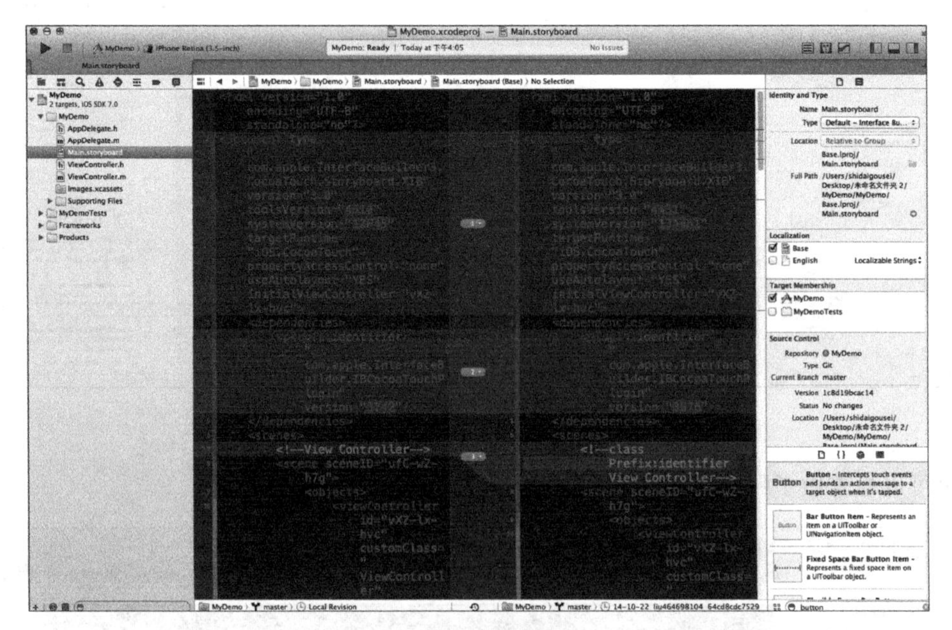

图 12.37 版本回退

(2) 对于个人开发者来说，Xcode 自带的版本控制工具基本上已经可以满足需求。对于团队开发者来说，还可能涉及代码服务端管理、分支的创建和管理等操作。

12.3 让用户升级

打开一个 App 时，如果这个 App 有新版本，一般情况下，App 会弹出一个对话框，提示用户版本更新及是否升级。当单击“是”时，系统就会跳转到该 App 在 App Store 中的下载页面。

下面介绍这个功能。

如果要检测 App 版本是否更新，那么需要获取手机端已安装的 App 版本号和 App 最新版本号。

(1) 获取手机端已安装的 APP 版本号，代码如下：

```
NSDictionary * infoDic= [[NSBundle mainBundle] infoDictionary];
NSString * appVersion= [infoDic objectForKey:@ "CFBundleVersion"];
```

(2) 获取 App 的最新版本号有两种方法。

① 在某特定的服务端上,发布和存储 App 的最新版本号,App 向该服务端请求查询。

② 从 App Store 上查询,可以获取 App 的作者、连接、版本等。

第 2 种方法不需要配置服务端,非常方便。下面介绍第 2 种方法。

发送请求访问:

http://itunes.apple.com/lookup? id=你的应用程序的 ID。

从获得的 response 数据中解析需要的数据。

从 App Store 查询得到的信息是 Json 格式,经过解析后得到的原始数据如下:

```
{
    resultCount=1;
    results=    (
                {
            artistId=开发者 ID;
            artistName=开发者名称;
            price=0;
            isGameCenterEnabled=0;
            kind=software;
            languageCodesISO2A=   (
                EN
            );
            trackCensoredName=审查名称;
            trackContentRating=评级;
            trackId=应用程序 ID;
            trackName=应用程序名称;
            trackViewUrl=应用程序介绍网址;
            userRatingCount=用户评级;
            userRatingCountForCurrentVersion=1;
            version=版本号;
            wrapperType=software;
        }
    );
}
```

从中取得 results 数组即可。latestVersion 就是 App 在 App Store 上的版本号。获取 latestVersion 代码如下:

```
NSDictionary * jsonData=[dataPayload JSONValue];
NSArray * infoArray=[jsonData objectForKey:@"results"];
NSDictionary * releaseInfo=[infoArray objectAtIndex:0];
NSString * latestVersion=[releaseInfo objectForKey:@"version"];
```

关键完整代码如下：

```
- (void)onCheckVersion
{
    NSDictionary * infoDic= [[NSBundle mainBundle] infoDictionary];
    NSString * currentVersion= [infoDic objectForKey:@"CFBundleVersion"];
    NSString * URL=@"http://itunes.apple.com/lookup?id=你的应用程序的 ID";
    NSMutableURLRequest * request= [[NSMutableURLRequest alloc] init];
    [request setURL:[NSURL URLWithString:URL]];
    [request setHTTPMethod:@"POST"];
    NSHTTPURLResponse * urlResponse=nil;
    NSError * error=nil;
    NSData * recervedData = [NSURLConnection sendSynchronousRequest:requ-
estreturningResponse:&urlResponse error:&error];
    NSString * results= [[NSString alloc] initWithBytes:[recervedData bytes]
length:[recervedData length] encoding:NSUTF8StringEncoding];
    NSDictionary * dic= [results JSONValue];
    NSArray * infoArray= [dic objectForKey:@"results"];
    if ([infoArray count]) {
        NSDictionary * releaseInfo= [infoArray objectAtIndex:0];
        NSString * lastVersion= [releaseInfo objectForKey:@"version"];
        if (![lastVersion isEqualToString:currentVersion]) {
            UIAlertView * alert= [[UIAlertView alloc] initWithTitle:@"更新"
message:@"有新的版本更新,是否前往更新?" delegate:self cancelButtonTitle:@"关
闭" otherButtonTitles:@"更新", nil];
            alert.tag=10000;
            [alert show];
        }
        else
        {
            UIAlertView * alert= [[UIAlertView alloc] initWithTitle:@"更新"
message: @" 此 版 本 为 最 新 版 本 " delegate: self cancelButtonTitle: @" 确 定 "
otherButtonTitles:nil, nil];
            alert.tag=10001;
            [alert show];
        }
    }
}
- ( void ) alertView: ( UIAlertView * ) alertView clickedButtonAtIndex:
(NSInteger)buttonIndex
{
    if (alertView.tag==10000) {
        if (buttonIndex==1) {
```

```
            NSURL * url= [NSURL URLWithString:@"https://itunes.apple.com"];
            [[UIApplication sharedApplication]openURL:url];
        }
    }
}
```

这里需要注意添加UIAlertView的协议：UIAlertViewDelegate。

12.4 基础知识与技能回顾

本章介绍了如何发布iOS应用。除本章介绍的方式外，开发者还可以申请企业发布证书，将应用打包后直接在设备上安装，不过这种方式不能再发布到App Store上。

本章还介绍了代码的版本管理。代码的版本管理非常重要，特别是多人协作的项目，版本管理能为开发者带来极大的方便。

练 习

项目

功能描述：用Xcode提交一次代码版本。

第 13 章 chapter 13

HTML 5

通过 HTML 5，开发者可以在一个代码库的基础上开发适用于不同设备的 App，其体验与原生 App 基本无异，不需要重复编程，也无需使用多重语言或 SDK。现代 Web 浏览器的发展使 HTML 5 能够实现跨平台、适用于不同设备的解决方案，这些方案与“原生”App 体验极为相似，往往很难分清它究竟是原生开发还是使用 HTML 开发。

13.1 什么是 HTML 5

HTML 5 是 HTML 下一个主要的修订版本，现在仍处于发展阶段。目标是取代 1999 年制定的 HTML 4.01 和 XHTML 1.0 标准，以期能在网际网络应用迅速发展的时候，使网络标准符合当代网络需求。广义论及 HTML 5 时，实际指的是包括 HTML、CSS 和 JavaScript 在内的一套技术组合。它希望能够减少浏览器对于需要外挂程序的丰富的网络应用服务(plug-in-based rich internet application，RIA)，如 Adobe Flash、Microsoft Silverlight 与 Oracle JavaFX 的需求，并且提供更多能有效增强网络应用的标准集。

HTML 5 提供了一些新的元素和属性，反映了典型的现代用法。其中有些在技术上类似于<div>和<span>标签，但有一定含义，例如<nav>(网站导航块)和<footer>。这种标签将有利于搜索引擎的索引整理、小屏幕装置和视障人士使用，同时为其他浏览要素提供了新的功能。

一些过时的 HTML 4 标记将取消，其中包括纯粹用作显示效果的标记，如<font>和<center>，因为它们已经被 CSS 取代。还有一些通过 DOM 的网络行为。

尽管和 SGML 在标记上相似，但 HTML 5 的句法已不再基于它了，而是被设计成向后兼容和对老版本 HTML 的解析。它有一个新的开始行，看起来就像 SGML 的文档类型声明<! DOCTYPE HTML>，这会触发和标准兼容的渲染模式。

13.2 用 HTML 5 实现内容展示

通常，iPhone 应用是利用 Cocoa Touch 框架以纯 Objective C 来构建的。可能会有一个 UITabBarController，上面安放了一些视图控制器(view controllers)。这些视图控制器可能是 UITableViewController 的子类，也可能是 UIViewController，其 UI 是使用 XIB 所定义或者在代码中定义。有时候，一个视图控制器上可能会安放一个 UIWebView 控件，用来在一个应用的内部展示 Web 内容或长文本内容。

混合应用是用 HTML、CSS 和 Javascript 而不是通过 Objective-C 来实现部分或所有的客户端代码。一个特定应用的屏幕实际上可能包含一个 UIWebView，用它来渲染服务端返回的标记语言（HTML）并且解释服务端返回的代码（JavaScript）而不是 Objective-C。

目前主流的一款混合开发框架是 PhoneGap。PhoneGap 是基于 HTML、CSS 和 JavaScript，用于创建移动跨平台应用程序的快速开发平台。它使开发者能够利用 iPhone、Android、Palm、Symbian、WP7、Bada 和 Blackberry 智能手机的核心功能，包括地理定位、加速器、联系人、声音和振动等。此外，PhoneGap 还拥有丰富的插件可以调用。

PhoneGap 的下载地址：

```
http://phonegap.com/download/
```

下面介绍如何利用 PhoneGap 进行 HTML 5 的开发。

（1）以 phonegap-1.8 版本为例。下载完成后，浏览至已解压文件的 lib/iOS/ 文件夹，双击 Cordova-1.8.0.dmg 包安装程序。其安装比较简单，直接单击 Next，直到安装完成。

（2）在 Xcode 中创建工程项目。单击 Xcode 欢迎屏幕上的 Create a new Xcode project，如图 13.1 所示。

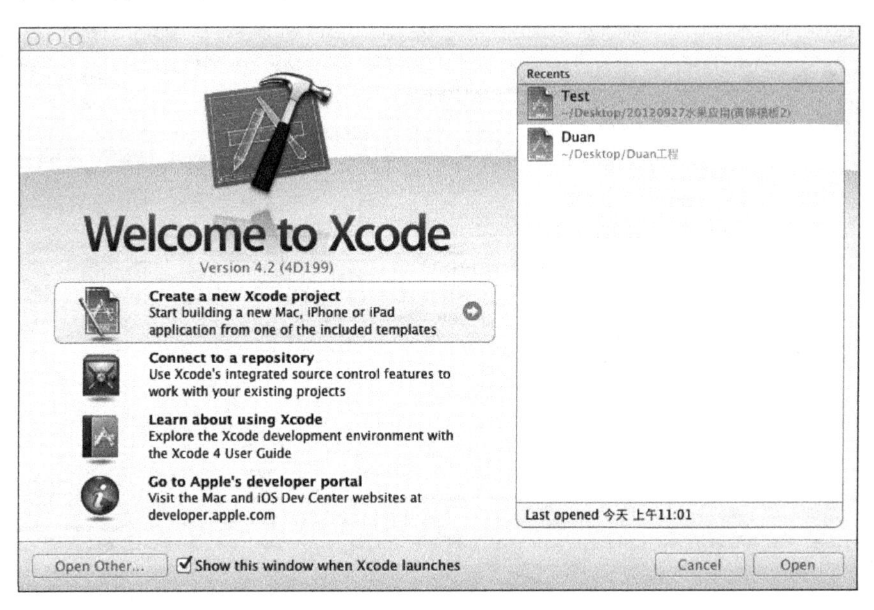

图 13.1　创建一个新项目

（3）在 iOS 的 Application 下找到并选中 PhoneGap-based Application 模板，然后单击 Next，如图 13.2 所示。

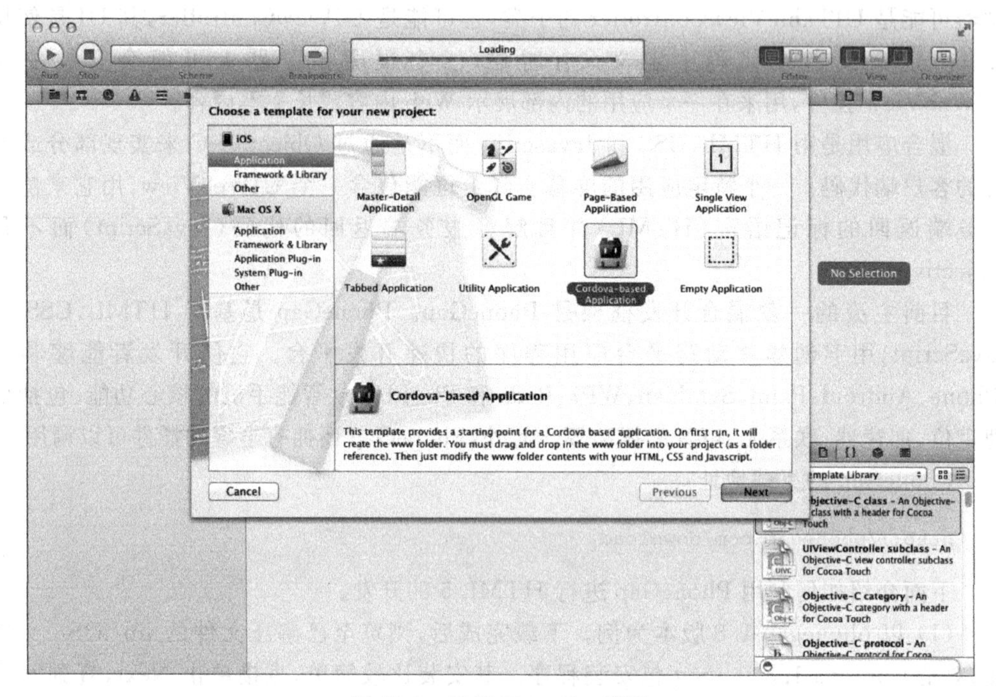

图 13.2　选择 PhoneGap 模板

(4) 输入工程项目名称，如图 13.3 所示。

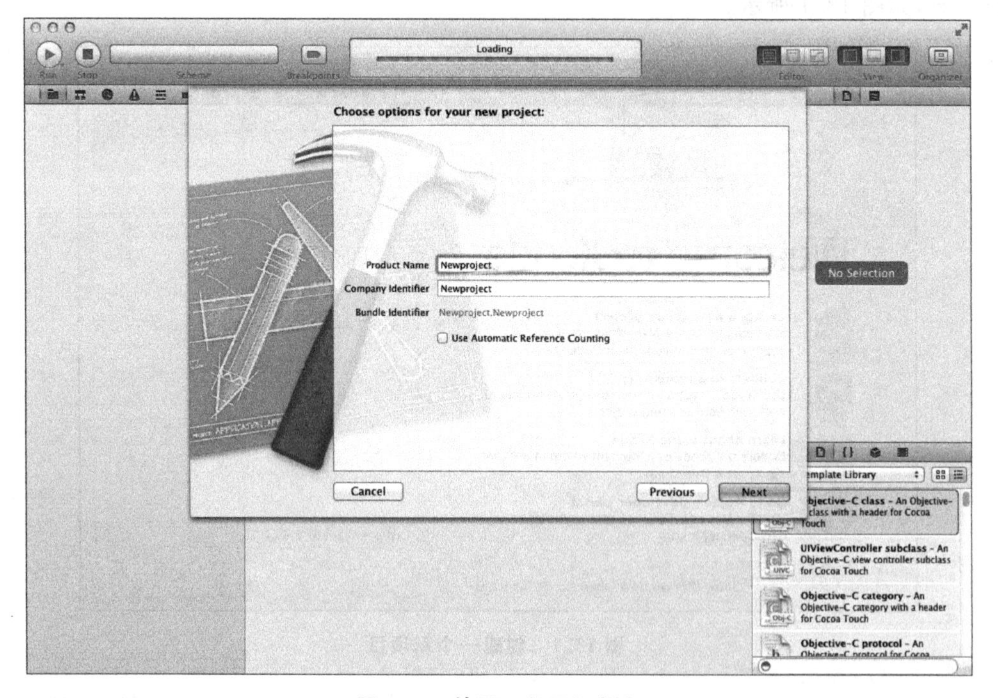

图 13.3　填写工程项目名称

(5) 选择存储路径后，按下 command+B 编译项目。编译完成后，项目文件夹就会多出一个名称为 www 的文件夹。右击项目名称，选择 Show in Finder 选项找到这个文件夹，如图 13.4 所示。

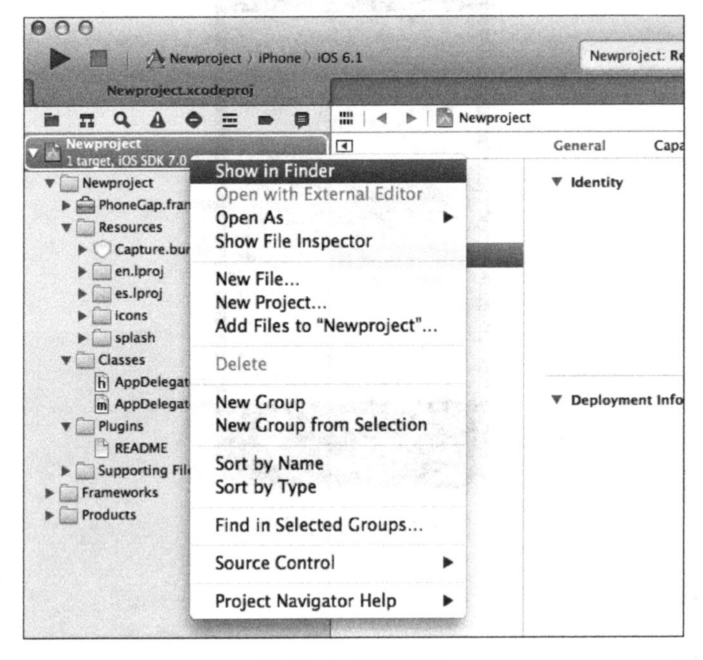

图 13.4 运行项目

(6) 将 www 文件夹添加至项目，如图 13.5 所示。

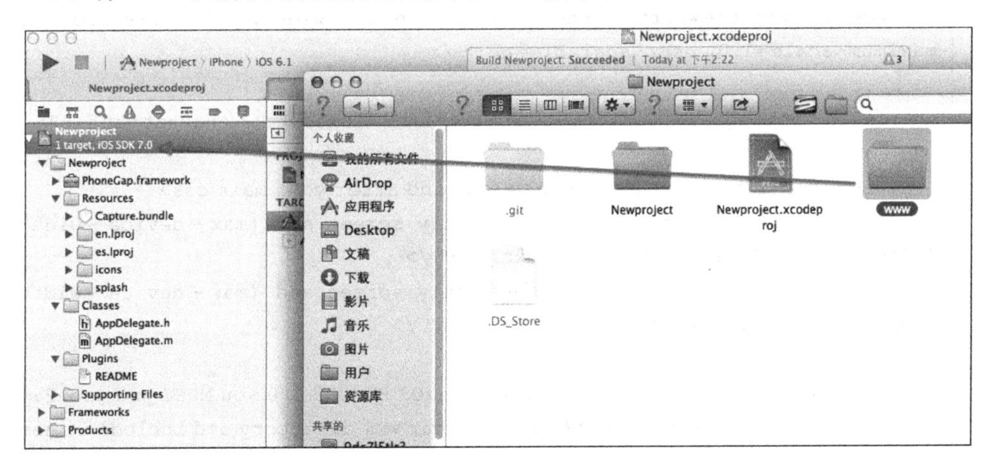

图 13.5 添加 www 文件夹

(7) 打开 www 文件夹下的 index. html 文件，这是运行项目时首先要执行的文件，而且可以针对不同的平台生成 App 文件。修改代码让模拟器打印“hello world”，如图 13.6 所示。

图 13.6　运行界面

代码如下：

```
<!DOCTYPE html>
<html>
  <head>
    <meta name="viewport" content="width=device-width, initial-scale=1.0,
maximum-scale=1.0, user-scalable=no;" />

    <meta http-equiv="Content-type" content="text/html; charset=utf-8">

    <!--iPad/iPhone specific css below, add after your main css>
    <link rel="stylesheet" media="only screen and (max-device-width:
1024px)" href="ipad.css" type="text/css" />
    <link rel="stylesheet" media="only screen and (max-device-width:
480px)" href="iphone.css" type="text/css" />
    -->
    <!--If your application is targeting iOS BEFORE 4.0 you MUST put json2.js
from http://www.JSON.org/json2.js into your www directory and include it here
-->
    <script type="text/javascript" charset="utf-8" src="phonegap-1.0.0.js">
</script>
    <script type="text/javascript" charset="utf-8">

    //If you want to prevent dragging, uncomment this section
    /*
    function preventBehavior(e)
```

```
    {
        e.preventDefault();
    };
    document.addEventListener("touchmove", preventBehavior, false);
    */

    /* If you are supporting your own protocol, the var invokeString will
contain any arguments to the app launch.
    see http://iphonedevelopertips.com/cocoa/launching-your-own-application-
via-a-custom-url-scheme.html
    for more details-jm */
    /*
    function handleOpenURL(url)
    {
        //TODO: do something with the url passed in.
    }
    */

    function onBodyLoad()
    {
        document.addEventListener("deviceready",onDeviceReady,false);
    }

    /* When this function is called, PhoneGap has been initialized and is
ready to roll */
    /* If you are supporting your own protocol, the var invokeString will
contain any arguments to the app launch.
    see http://iphonedevelopertips. com/cocoa/launching - your - own -
application-via-a-custom-url-scheme.html
    for more details-jm */
    function onDeviceReady()
    {
        //do your thing!

        alert("hello world")
    }

    </script>
  </head>
  <body onload="onBodyLoad()">
      <h1>Hey, it's PhoneGap!</h1>
      <p>Don't know how to get started? Check out<em><a href="http://github.
com/phonegap/phonegap-start">PhoneGap Start</a></em>
  </body>
</html>
```

13.3 基础知识与技能回顾

本章介绍了如何用 PhoneGap 结合 HTML 5 的方式开发 iOS 应用。利用 PhoneGap 框架,开发者能够方便、简单、快速地进行混合开发。感兴趣的读者可以详细了解这种开发模式。

练　　习

项目

功能描述：搭建 PhoneGap 开发环境并用模拟器打印"hello world"。

参 考 文 献

1. Bob LeVitus. Incredible iPhone Apps FOR Dummies[M]. Wiley Publishing, Inc. 2010.

2. Dave Mark, JeffLaMarche. More iPhone 3 Development Tackling iPhone SDK 3[M]. Appress, 2009.

3. Dave Mark, Jeff LaMarche. iPhone 3 开发基础教程[M]. 漆振, 杨越, 孙文磊译. 北京：人民邮电出版社, 2009.